An international initiative by the
World Federation for Culture Collections,
with financial support from UNESCO

LIVING RESOURCES FOR BIOTECHNOLOGY

Filamentous Fungi

LIVING RESOURCES FOR BIOTECHNOLOGY

Filamentous Fungi

Edited by

D. L. Hawksworth and B. E. Kirsop

in collaboration with

S. C. Jong, J. I. Pitt, R. A. Samson and K. Tubaki

The right of the
University of Cambridge
to print and sell
all manner of books
was granted by
Henry VIII in 1534.
The University has printed
and published continuously
since 1584.

CAMBRIDGE UNIVERSITY PRESS
Cambridge
New York   New Rochelle   Melbourne   Sydney

Published by the Press Syndicate of the University of Cambridge
The Pitt Building, Trumpington Street, Cambridge CB2 1RP
32 East 57th Street, New York, NY 10022, USA
10 Stamford Road, Oakleigh, Melbourne 3166, Australia

First published 1988

Printed in Great Britain at the University Press, Cambridge

British Library cataloguing in publication data
Filamentous fungi. – (Living resources for
biotechnology).
1. Fungi
I. Hawksworth, D. L.   II. Kirsop, B. E.
III. Series
589.2   QK603

Library of Congress cataloguing in publication data
Filamentous fungi – edited by D. L. Hawksworth and B. E. Kirsop in
collaboration with S. C. Jong . . . [et al.].
p. cm. – (Living resources for biotechnology)
Includes index.
ISBN 0–521–35226–6
1. Fungi – Biotechnology.   I. Hawksworth, D. L.   II. Kirsop, B. E.
III. Jong, S.   IV. Series.
TP248.27.F86F55 1988
660'.62–dc19   87–27721   CIP

ISBN 0 521 35226 6

CONTENTS

CONTRIBUTORS

Allner, K. Public Health Laboratory Service Centre for Applied Microbiology and Research, Porton Down, Salisbury, Wiltshire SP4 0JG, UK (Chapter 3)

Allsopp, D. CAB International Mycological Institute, Ferry Lane, Kew, Surrey TW9 3AF, UK (Chapter 7)

Bousfield, I. J. National Collections of Industrial and Marine Bacteria Ltd, Torry Research Station, 135 Abbey Road, PO Box 31, Aberdeen AB9 8DG, UK (Chapter 6)

DaSilva, E. J. Division of Scientific Research and Higher Education, United Nations Educational Scientific and Cultural Organisation, 7 Place de Fontenoy, 75700 Paris, France (Chapter 8)

Fabricius, B.-O. Department of Microbiology, University of Helsinki, SF-00710 Helsinki, Finland (Chapter 2)

Hawksworth, D. L. CAB International Mycological Institute, Ferry Lane, Kew, Surrey TW9 3AF, UK (Chapters 1, 3, 5)

Jong, S. C. American Type Culture Collection, 12301 Parklawn Drive, Rockville, Maryland 20852, USA

Kirsop, B. E. Microbial Strain Data Network, Institute of Biotechnology, University of Cambridge, 307 Huntingdon Road, Cambridge CB3 0JX, UK (Chapter 8)

Krichevsky, M. I. Microbial Systematics Section, Epidemiology and Oral Disease Prevention Program, National Institute of Dental Research, Bethesda, Maryland 20892, USA (Chapter 2)

Onions, A. H. S.   CAB International Mycological Institute, Ferry Lane, Kew, Surrey TW9 3AF, UK (Appendix)

Pitt, J. I.   Division of Food Research, CSIRO, PO Box 52, North Ryde, New South Wales 2113, Australia (Appendix)

Samson, R. A.   Centraalbureau voor Schimmelcultures, P.O. Box 273, 3740 AG Baarn, The Netherlands

Simione, F. P.   American Type Culture Collection, 12301 Parklawn Drive, Rockville, Maryland 20852, USA (Chapter 7)

Smith, D.   CAB International Mycological Institute, Ferry Lane, Kew, Surrey TW9 3AF, UK (Chapter 4)

Sugawara, H.   Life Science Research Information Section, RIKEN, Wako, Saitama 351-01, Japan (Chapter 2)

Tubaki, K.   University of Tsukuba, Institute of Biological Sciences, Sakura-mura, Ibaraki 305, Japan

SERIES INTRODUCTION

The rapid advances taking place in biotechnology have introduced large numbers of scientists and engineers to the need for handling microorganisms, often for the first time. Questions are frequently raised concerning sources of cultures, location of strains with particular properties, requirements for handling the cultures, preservation and identification methods, regulations for shipping, or for the deposit of strains for patent purposes. For those in industry, research institutes or universities with little experience in these areas, resolving such difficulties may seem overwhelming. The purpose of the World Federation for Culture Collections' (WFCC) series, Living Resources for Biotechnology, is to provide answers to these questions.

Living Resources for Biotechnology is a series of practical books that provide primary data and guides to sources for further information on matters relating to the location and use of different kinds of biological material of interest to biotechnologists. A deliberate decision was taken to produce separate volumes for each group of microorganism rather than a combined compendium, since our enquiries suggested that inexpensive specialised books would be of more general value than a larger volume containing information irrelevant to workers with interests in one particular type of organism. As a result each volume contains specialised information together with material on general matters (information centres, patents, consumer services, the international coordination of culture collection activities) that is common to each.

The WFCC is an international organisation concerned with the establishment of microbial resource centres and the promotion of their activities. In addition to its primary role of coordinating the work of culture collections throughout the world, the committees of the WFCC are active in a number of areas of particular relevance to biotechnology, such as patents, microbial information centres, postal and quarantine regulations, educational and conservation matters (see Chapter 8). The Education Committee of the WFCC proposed the preparation of the current volumes.

The WFCC is concerned that this series of books is of value to biotechnologists internationally, and the authors have been drawn from specialists throughout the world. The close collaboration that exists between culture collections in every continent has made the compilation of material for the books a simple and pleasurable process, since the authors and contributors are for the most part colleagues. The Federation hopes that the result of their labours has produced valuable source books that will not only accelerate the progress of biotechnology, but will also increase communication between culture collections and their users to the benefit of both.

Barbara Kirsop
President, World Federation
for Culture Collections

PREFACE

The filamentous fungi represent the group of microorganisms with the largest number of species, showing an immense variety not only in morphology but also in physiological and biochemical attributes. About 63 700 species (excluding yeasts) are currently known, but around 1500 are described as new to science each year, and the number in nature may well exceed 250 000. Over 3000 secondary metabolites have already been characterised, but actual numbers are far in excess of this and the biological activities of most remain to be determined. The biotechnological importance of the filamentous fungi is, therefore, of considerable significance, and the potential of this vast resource is only now starting to be appreciated by biotechnologists. Fortunately they have at their disposal over 170 000 strains maintained in culture collections throughout the world.

This book provides an introduction to these resources and describes how information can be obtained on what is available, how filamentous fungi can be preserved and identified, how collections operate, and the additional support services available from them.

In preparing this volume, I have been fortunate in securing the assistance of colleagues from different parts of the world who are experienced with the work of culture collections, as curators or users, as well as from the individual collections themselves. Without their help the international overview this text aims to provide could not have been presented.

Kew
January 1987

D. L. Hawksworth
Director
CAB International Mycological Institute

ACKNOWLEDGEMENTS

This book represents a collaborative effort. Drafts for each chapter were prepared and then circulated to Dr S. C. Jong (American Type Culture Collection, USA), Dr J. I. Pitt (CSIRO Division of Food Research, Australia), Dr R. A. Samson (Centraalbureau voor Schimmelcultures, The Netherlands) and Professor K. Tubaki (University of Tsukuba, Japan), whose comments have been incorporated into the final version.

Authors of individual chapters were: Chapter 2, Dr M. Krichevsky (National Institute of Health, USA), B.-O. Fabricus (University of Helsinki, Finland) and H. Sugawara (RIKEN, Japan); Chapter 4, Dr D. Smith (CAB International Mycological Institute, UK); Chapter 6, Dr I. Bousfield (National Collections of Industrial and Marine Bacteria Ltd, UK); Chapter 7, Dr D. Allsopp (CAB International Mycological Institute, UK) and Dr F. P. Simione (American Type Culture Collection, USA); and Chapter 8, Mrs B. Kirsop (Microbial Strain Data Network, UK) and Dr E. J. DaSilva (UNESCO, Paris). Dr K. Allner (PHLS Centre for Applied Microbiology and Research, UK) prepared a draft of part of the section on safety regulations in Chapter 3, and Dr A. H. S. Onions (CAB International Mycological Institute, UK) and Dr J. I. Pitt (CSIRO, Australia) were largely responsible for the Appendix. The remaining drafts were prepared by the volume co-ordinator (D. L. H.).

The co-ordinator is also grateful to Dr A. H. S. Onions (CAB International Mycological Institute) for reading and commenting on a complete draft of the work.

The details on individual collections incorporated into Chapter 1, Section 1.5 were compiled from drafts prepared by the collections themselves specially for this volume.

1

Resource centres

D. L. HAWKSWORTH

1.1 Nature of the resource

Although culture collections of filamentous fungi date back to the late nineteenth century and one of the largest of them, the Centraalbureau voor Schimmelcultures (CBS) in The Netherlands was established in 1904, rather little interest had been shown in their funding, and proposals to set up such collections often received only token support. With the advent of biotechnology the search for microorganisms which have properties suitable for commercial exploitation has led to a renewed interest in culture collections because of the appreciation of the valuable resource they represent.

The term 'filamentous fungi' is used for species producing filament-like hyphae. It includes almost the entire fungal kingdom and is used in contradistinction to 'yeasts', which are essentially unicellular fungi with vegetative cells capable of repeated budding. The yeast fungi are not considered further here as they are treated in a companion volume in this series. While some fungus culture collections maintain both filamentous fungi and yeasts, most specialise in one growth form or the other. This is a consequence of the different uses made of them (and so of their relevance to particular industries), the diverse growth conditions, and the dissimilar ways in which they are currently characterised (physiologically and biochemically in the yeasts, but largely morphologically in all other fungi).

Precisely how many strains of filamentous fungi are maintained in the living state in culture collections throughout the world is unknown. However, the number certainly exceeds 170 000 scattered through over 200 collections, and it has been estimated that around 7000 different species are represented (Hawksworth, 1985a). The physiological and

1

biochemical attributes of the majority of these strains have not yet been determined. They constitute a vast and as yet largely unexploited resource, and it is for this reason that workers in various areas of biotechnology and related industries have begun to take such interest in culture collections in the 1980s.

Culture collections are, however, much more than places for the deposit and supply of cultures. They are almost always managed by specialist mycologists and thus form part of larger centres of mycological expertise. The resource centres consequently provide expert advice on filamentous fungi and their properties, their identification and preservation. In addition, training and specialist services to industry on a contract or consultancy basis are available from the larger collections, a number of which also act as depositories for patent strains. The aim of this book is to describe the nature of the resources and services available from culture collections of filamentous fungi for the benefit of biotechnologists and other potential users.

1.2   Biotechnological applications

The oldest established uses of filamentous fungi are those concerned with food for man. Larger fleshy fungi (particularly mushrooms) collected from pastures and forests have been eaten by man from the earliest times. The number of edible species probably exceeds 500, but many are restricted to particular geographical regions. Remarkably, not more than 20 species are currently exploited commercially (e.g. *Agaricus brunnescens*, 'mushrooms'; *Lentinula edodes*, 'shii-take'; *Tricholoma matsutake*, 'matsu-take'; *Volvariella volvacea*, 'padi straw mushroom'). Opportunities clearly exist for developing techniques to extend the numbers of species cultivated commercially. These could be of considerable value as a supplementary food source in less-developed countries. An important review of the cultivation of edible mushrooms is given by Chang & Hayes (1978), and of their sources by Wu (1987).

Many fermented foodstuffs and drinks, especially in Asia, are based on filamentous fungi (Hesseltine, 1965; Batra & Millner, 1976; Steinkraus, 1983). The most important of these are species of *Aspergillus* (e.g. *A. oryzae*, Koji), *Monascus* (e.g. *M. purpureus*, 'red rice', ang-kak), *Mucor* (e.g. *M. rouxii*, 'Chinese rice') and *Rhizopus* (e.g. *R. oligosporus*, 'tempeh'). The role of certain *Penicillium* species, such as *P. camembertii* and *P. roquefortii*, in cheese production also has an extremely long history (Pitt, 1980).

These long-established uses in food production are being extended by biotechnologists to the production of single-cell protein (SCP) for use in both human foods and animal feeds (Birch, Parker & Worgan, 1976; Smith, 1981). The filamentous fungi are grown in large fermenters with appropriate nutrients, which may be waste materials such as wood chips. *Fusarium*-based products (Anderson & Solomons, 1984) are now on sale for human consumption in the UK, and other genera with potentially useful species include *Chaetomium*, *Paecilomyces* and *Trichoderma*. In these applications, strain selection is of special importance, as some species of all these genera produce harmful mycotoxins. Where biodegradation of wastes rather than production of feedstuffs is the desired end-product of the process, toxin production may not be a significant consideration. Various cellulases and lignase enzymes are produced by conidial and mycelial phases of wood-rotting Hymenomycetes, such as the *Sporotrichum* species.

Fungi also have applications in the detoxification of wastes, and in the extraction of metals from dilute solutions (Eggins & Allsopp, 1975).

Filamentous fungi are a rich source of metabolites. Over 3000 have been characterised, and novel compounds continue to be discovered (Turner, 1971; Turner & Aldridge, 1983). Some 1600 of these have already been found to have antibiotic or antitumour properties. The biological activities of many fungi are still unknown, however, and representatives of only a small proportion of known genera have yet been screened. The cephalosporins and penicillins are of vital importance as antibiotics. Other important products are the anti-fungal griseofulvin (from *Penicillium griseofulvum*), ergot alkaloids (from *Claviceps purpurea*), plant growth hormones (gibberellins, from *Fusarium moniliforme*), and the cattle growth stimulator zearalenone (from *F. graminearum*). Exciting attributes are still coming to light. For example, the immunosuppressant cyclosporin from *Tolypocladium inflatum* was only discovered in 1970, but is now transforming the safety of transplant surgery, and the cytocalasins from *Hypoxylon* and related species found in 1967 are proving of value in isolating nuclei for genetic engineering.

Fungi are also sources of industrial chemicals. Examples of these are citric and gluconic acids (*Aspergillus niger*), vitamins (riboflavin from *Eremothecium ashbyii*), polysaccharides (pullulan from *Aureobasidium pullulans*), and enzymes (rennin from *Rhizomucor pusillus*, lipase from *Penicillium roquefortii*, protease from *Aspergillus oryzae*, cellulase from *Trichoderma viride*). A wide variety of biochemical conversions and modifications of molecules such as sterols can also be carried out by fungi. Bennett (1985) and Onions, Allsopp & Eggins (1981) provide useful introductions to the diverse industrial uses of fungi. As the activity of strains maintained in culture collections becomes better documented, potential new sources of valuable compounds will be identified.

Particularly exciting are the possibilities for genetic engineering using filamentous fungi (Bennett & Lasure, 1985). These are only just starting to be explored, but somatic hybridisation between allied species and the potential for the transfer into bacteria of genes making valuable metabolites present many new possibilities. An enzyme from *Trichoderma* has recently been cloned and expressed in yeast (Ardsell *et al.*, 1987).

With world concern over the possible harmful effects of agricultural pesticides, increasing emphasis is being placed on the use of fungi as both mycoherbicides (strains of *Colletotrichum gloeosporioides*) and mycoinsecticides (species of *Beauveria, Hirsutella, Metarhizium, Verticillium*). Several products based on the mass inoculation of targets by conidia are now commercially available.

Increased precision in the identification at the strain level of mycorrhizal fungi associated with both forest trees and orchids, and the use of biotechnological methods to improve the effectiveness of strains, have considerable promise for plantation forestry productivity and the orchid industry.

1.3   Conservation

Culture collections may be considered as germ plasm resources equivalent to seed banks, botanical gardens and zoos, in that they conserve representatives of the world's microorganisms and make them available to others for propagation and use. The maintenance of a fully comprehensive range of strains of the world's filamentous fungi is a daunting task never likely to be achieved by a single culture collection. In general, collections hold strains relevant to the interests of past and present staff scientists and to the applied field of the funding body, or to meet the needs of the region in which they are located. Most strains in individual collections are thus not replicated in other collections. As a result of the different specialisations of collections, they may be seen as components of a collective resource.

By using this resource, especially when assisted by the computerised information systems now being developed (Chapter 2), biotechnologists are able to obtain strains from a wide variety of substrata and regions without having to mount expensive collecting expeditions. Furthermore, some filamentous fungi are extremely rare in nature and their rediscovery by isolation from soils or other materials may be far from easily accomplished. A considerable number of fungi described in culture are only known from the original isolation.

Culture collections have an obligation to conserve this massive resource for posterity and are engaged in developing increasingly successful long-term methods to attain this objective. Confidential short-term conservation of cultures for research workers in industry and universities is also undertaken by culture collections which provide a 'safe-deposit' facility (Chapter 7). This service reduces the risk of loss due to contamination or accident during experimental investigations and also limits the possibility of genetic change that can arise from inappropriate or poor preservation procedures, such as periodic transfer used in many research laboratories.

The value of culture collections is enhanced both by the acquisition of new isolates through the work of collection staff and by depositors submitting isolates. The policy on acquisitions varies considerably between collections (Chapter 3, Section 3.2.1) and some are not in a position to accept all cultures offered to them, because of limited staff and facilities. However, all research workers are urged to ensure that living cultures on which published research results are based are permanently conserved in a service culture collection. If this is not done, future workers may not be able to extend or verify the results, in the latter case rendering the publication of doubtful value. Much elegant biochemical work on filamentous fungi has already been based on strains which are not now available for further study.

1.4 Types of culture collections

Existing culture collections were established with a variety of objectives and priorities to fulfil different functions. Collections may generally be regarded as belonging to one of the following four categories, but it must be stressed that these are not mutually exclusive as some of the larger collections operate in more than one mode.

1.4.1 *Service collections*

Service culture collections have as their main objective the provision of authenticated cultures to all who request them.

Most service collections maintain a broad spectrum of fungi (and often other groups of organisms as well), actively solicit particular strains cited in the literature, and are concerned with the development of long-term preservation methods (Chapter 4). Many are also International Depositary Authorities (IDAs) for patent strains (Chapter 6). While the income from the sale of cultures by the largest service collections is not inconsiderable, many strains are retained for conservation purposes and are rarely sold. Collections cannot recover their full costs from such sales. Consequently, service collections are generally supported by substantial grants from public funds in recognition of the fundamental service they provide to pure and applied biology.

The ability of service collections to check the identity of strains both on deposit and during the course of maintenance and preservation procedures is vital to a collection's ability to supply authenticated strains. Many collections have expert taxonomic mycologists on staff, or are associated with institutions with such scientists, in order to be confident of this aspect of quality control.

The major service collections produce catalogues of the strains held, are involved in advisory and training work to varying extents, conduct research on systematics and preservation, and often undertake identification work. Some also have facilities for biochemical, biodeterioration, or other specialised work and these are listed in Section 1.5 below.

1.4.2   *In-house collections*

In-house collections are those established primarily to serve the needs of the organisation of which they form a part, whether it be a government department, research laboratory or an industrial company. In most cases, in-house collections established in publicly funded institutes supply cultures to others, but only as a secondary function and subject to them having the manpower to do so. Catalogues are rarely available, charging may or may not be implemented, and other services are related to the activities of the parent body. Several major industrial companies have very large in-house collections, and highly sensitive strains used in commercial operations or the subject of active research are kept under strict security. However, some companies make isolates available from time to time to other researchers when this will not endanger commercial operations. As catalogues are seldom available to the public, the use of such in-house collections by others is necessarily limited.

1.4.3   *Research collections*

Individual research workers regularly build up substantial collections of cultures related to their own research work. These collections are often of particular importance as they contain strains with unusual attributes that may be cited in publications by the researcher. Such collections are often maintained only by short-term maintenance procedures (repeated transfers, domestic deep-freeze) and the parent institution may have no commitment to their long-term preservation. There are consequently risks of loss due to the use of inadequate preservation methods, and the collection can become endangered when the research worker changes employment or research interests, completes a higher degree or retires.

It is very important that research workers deposit subcultures of key isolates with major service collections for long-term preservation. Collections can hold them in 'Reserve' or as 'Safe-Deposits' (Chapter 7, Section 7.2.1) if the researcher does not want them to be generally available until his or her work is completed and published, but in the meantime they will be expertly maintained.

Many research workers make available selected isolates to colleagues and others working in the same field on a free or exchange basis.

1.4.4   *Laboratory suppliers*

Several commercial laboratory suppliers include a few filamentous fungi among the lists of living organisms they have available for sale. These are usually single representatives of species widely used for teaching purposes at schools or at undergraduate level. Prices are usually competitive with those of service collections due to the large-volume sales of a very few strains. In the case of the filamentous fungi, strains maintained by such suppliers are limited to the extent that they are likely to be of little value to research workers in biotechnology and related fields.

1.5   Resource centres

1.5.1   *Locating resource centres*

The number of culture collections that maintain strains of filamentous fungi is certainly in excess of 200, and many of these are included in the second edition of the *World Directory of Collections of Cultures of Microorganisms* (McGowan & Skerman, 1982) [available from The Secretary, UNEP/UNESCO/ICRO Panel on Microbiology, Swedish University of Agricultural Sciences, S-750 07 Uppsala, Sweden (book form or in microfiche), and also from UNEP]. These collections vary greatly in coverage, size and the level of services they provide. In the following sections, brief details are given on collections holding more than 500 strains, with extended treatment of those with over 10 000 strains. The information given here has been supplied especially for this publication by the collections; only those with over 500 strains and that wished their work to be listed are included. Information on further collections, and additional data on those listed is provided by McGowan & Skerman (1982), Hawksworth (1985a) and, for European collections, by the European Culture Collections' Organisation (1984).

Locating the resource centre, or centres, most likely to be able to provide particular strains or services is not always an easy task. Although the species held in many centres are listed in the *World Directory*, the details of strains are not included. Computerised databases play a key role in alleviating the present problem (Chapter 2), but the service collections listed below may also be contacted for advice on the location of particular strains. Additionally, the international or national organisations concerned with culture collections, microbiology, or biotechnology may be approached, and further information on these is provided in Chapter 8.

The acronyms adopted below are those used in the *Directory* of the World Data Center.

1.5.2   Asia

Culture Collection Department   CCCCM
Institute of Microbiology
Academia Sinica (AS), Beijing, China
Tel.: 28-5614
Founded: 1951   *Strains of fungi:* 4750
Status: Under the leadership of the China Committee for Culture Collections of Microorganisms. Since 1979 this collection has acted as the Centre for General Microbiological Culture Collections (CGMCC) in the Committee.
Coverage: All groups. Collection, preservation and distribution of cultures of microorganisms with significance to industry, taxonomy, biochemistry, genetics and education.
Notes: Catalogue available.

Institute of Applied Microbiology   IAM
The University of Tokyo, 1-1-1 Yayoi
Bunkyo-ku, Tokyo 113, Japan
Founded: 1953   *Strains of fungi:* 1000

Coverage: All groups.

Catalogue: Strains are listed in the Japan Federation for Culture Collections (JFCC) *Catalogue of Cultures* (available from the Business Center for Academic Societies Japan, 16-3 Hongo 6-chome, Bunkyo-ku, Tokyo 113). An independent catalogue is published.

Notes: The collection was established to promote research in basic and applied microbiology by collecting, preserving and distributing authentic cultures. This is one of the main purposes for which the Institute was founded. The collection is a member of the Japan Federation for Culture Collections. Cultures are distributed to institutions, both in Japan and abroad. Preservation is by freeze-drying, L-drying, cryogenic freezing, and serial transfer. Deposits of cultures of scientific interest are welcomed.

Institute for Fermentation   IFO
17-85 Juso-honmachi
2-chome, Yodogawa-ku, Osaka 532, Japan
Tel.: 06-302-7281
Founded: 1944   *Strains of fungi:* 7000
Status: Financial support is obtained from the IFO foundation and annual donations from a private company.

Coverage: All groups.

Catalogue: The IFO *List of Cultures* (7th edition, 1984) is available from the Business Center for Academic Societies Japan, 16-3 Hongo 6-chome, Bunkyo-ku, Tokyo 113, Japan.

Japan Collection of Microorganisms   JCM
The Institute of Physical and Chemical Research (RIKEN),
Wako
Saitama 351-01, Japan
Tel.: 0484-62-1111 ext. 6000; Telex: 02962818 (RIKENJ)
Founded: 1980   *Strains of fungi:* 1000
Status: Government supported.

Coverage: All groups.

Notes: Catalogue available. The foundation of the collection was stimulated by national policy on the promotion of life sciences. The collection is expected to function as the key centre in Japan. It collects, preserves and distributes moulds of scientific, industrial, medical and ecological importance, in active cooperation with other collections. Also, since it shares the host institute, RIKEN, JCM contributes to the activities of the World Data Center on Microorganisms (WDC) (see pp. 49–51).

1.5.3   *Australasia*

Australian National Reference Laboratory in Medical Mycology   AMMRL
Royal North Shore Hospital of Sydney
Pacific Highway, St Leonards
New South Wales 2065, Australia
Tel.: (02) 438-7128
Founded: 1949   *Strains of fungi:* 750
Coverage: All groups of medical interest except Basidiomycotina and Mastigomycotina. Primarily concerned with the maintenance and distribution of fungi of medical importance.
Catalogue: A catalogue is available.
Special services: The collection provides information on and assistance with diagnostic and technical problems in the field of medical mycology.
Research: Engaged in research which is directed towards the study of pathogenic fungi in Australia.
Training: Informal training programme for those working in the area of medical microbiology.

Victoria University of Wellington   DBVU
Microorganism Culture Collection
School of Biological Sciences, Private Bag
Wellington, New Zealand
Tel.: (64) (04) 721-000; Telex: NZ 30882
Founded: 1965   *Strains of fungi:* 594
Coverage: All groups. The collection contains fungi isolated from diseased plant material, cereal seeds, soil, jet and diesel fuel, bird feathers and air. Mercury resistant strains of *Pyrenophora avenae* and triadimenol and nuarimel resistant strains of *P. teres* and *P. graminea* are held.
Catalogue: A catalogue is available.
Culture sales: Cultures are exchanged free of charge for cultures from other collections and generally free to teaching institutions. A fee is charged to industry.
Preservation: Cultures are stored under sterile mineral oil.

Food Research (CSIRO) FRR
PO Box 52, North Ryde
New South Wales 2113, Australia
Tel.: (02) 887 8333; Telex: 23407
Founded: 1969   *Strains of fungi:* 3500
Coverage: All groups except Basidiomycotina and Mastigomycotina; principally food spoilage, xerophilic and mycotoxin-producing fungi.
Catalogue: No catalogue is available.
Notes: Mainly services the Australian food industry. Maintains the largest collection of xerophilic fungi; virtually all ex-type strains of *Penicillium* species known in culture are held. Cultures available on request; industrial users charged. Identification of food spoilage and xerophilic fungi may be carried out by prior arrangement.

Materials Research Laboratories MRL
PO Box 50, Ascot Vale, Victoria 3032
Australia
Tel.: 319-4499; Telex: 35230; Facsimile: 318-4536
Founded: 1940   *Strains of fungi:* 1000
Status: Commonwealth of Australia, Department of Defence, Defence Science and Technology Organization.
Coverage: Mostly saprophytic deuteromycetes. The earliest isolates date from 1940 when the establishment first became interested in biodeterioration of materials; this has remained the central theme. Geographically the isolates are mostly from within Australia, especially from Victoria and Northern Queensland and were derived from utilitarian materials and the environment. Only a few have been lodged with the larger international collections, and the biochemical characteristics of the strains remain largely uninvestigated. The collection includes significant sections of *Curvularia* and *Rhinocladiella sensu lato*.
Notes: Mould growth testing.

International Collection of Micro-organisms from Plants PDDCC
Mount Albert Research Centre, DSIR
Private Bag, Auckland, New Zealand
Tel.: Auckland 893-660; Telex: MARC NZ 21623
Founded: 1961   *Strains of fungi:* 2400
Status: The collection is the property of the Plant Disease Division, Department of Scientific and Industrial Research. Formerly the Plant Diseases Division Culture Collection (to 1987).

Coverage: 'Phycomycetes', Ascomycotina and Deuteromycotina. The collection is primarily of plant pathogens and related organisms. The fungal collection, mainly of strains from localities in New Zealand, exists to maintain cultures of plant pathogenic and associated fungi for use by research and educational establishments and by industry. It is intended that sufficient cultures of each species be kept to be representative of its geographic and host range, and of the variation within it.

Catalogue: A catalogue is available.

Culture sales: A charging policy for ampoules has recently been introduced. The Curator welcomes strains from other countries, and will waive charges in exchange.

Preservation: Most strains are held as lyophilized hyphal cultures. A small number are maintained under oil.

Plant Pathology Branch Culture Collection   QA
Department of Primary Industries
Meiers Road, Indooroopilly, Queensland 4068
Australia
Tel.: (07) 3779311; Telex: AA 41620; Facsimile (07) 8703276
                                                      Strains of fungi: 600
Coverage: Agricultural, principally plant pathogenic and plant associated microfungi. All groups except Basidiomycotina.

Catalogue: No independent catalogue available.

Culture sales: Cultures available free.

Notes: Associated with the herbarium of the Plant Pathology Branch (BRIP) at the same address.

Plant Research Division Culture Collection   WA
Plant Pathology Branch
Department of Agriculture, Baron-Hay Court
South Perth, Western Australia 6151
Tel.: (09) 368-3333; Telegrams: AGDEP Perth;
Telex: AA 93304
Founded: 1954   *Strains of fungi:* 5000
Status: Sponsored by the State Government.

Coverage: The collection comprises mostly fungi associated with plant diseases in Western Australia.

Catalogue: Details about the specimens are currently being put in a computer database.

Culture sales: Specimens are freely exchanged and distributed on request.

Preservation: Mostly held as lyophilized cultures.

Special services: Primarily used for reference purposes.

1.5.4 Europe

Aquatic Phycomycete Culture Collection   AP
Department of Botany, Plant Science Laboratories
University of Reading, Whiteknights
PO Box 221, Reading RG6 2AS, UK
Tel.: 0734-875123; Telex: 847813
Founded: 1956   *Strains of fungi:* 800
Status: Sponsored by the Department of Botany.

Coverage: Primarily saprophytic and parasitic Oomycetes, with a few other aquatic fungi. Most of the strains held are referrable to the families Saprolegniaceae and Pythiaceae, and many of these have been isolated by Dr M. W. Dick and his co-workers.

Catalogue: A catalogue will be published as a subset of the IMI catalogue.

Notes: Integration with IMI (see below) is in hand.

Centraalbureau voor Schimmelcultures   CBS
PO Box 273
3740 AG Baarn, The Netherlands
Tel.: 2154-11841
Founded: 1904   *Strains of fungi:* 29 500
Status: Institute of the Royal Netherlands Academy of Arts and Science (Amsterdam); Centre of the Netherlands Culture Collections and Dutch Microbial Network. Supported also by the Dutch Program Bureau Biotechnology. An institute of the University of Utrecht. See de Hoog (1979).

Coverage: All groups, especially isolates of industrial, agricultural, medical, forestry, biodeterioration, food microbiology and systematic importance.

Catalogue: Hard copy or microfiche revised every 3–4 years. New accessions published twice a year in newsletter. Data are computerised and distributed over 132 fields and available on-line. The Netherlands Node for MINE (see Chapter 2).

Culture sales: World-wide. Reductions for non-commercial use.

Sets for biodeterioration tests and educational purposes. Exchange considered.

Preservation: Agar slants, freeze-drying, liquid nitrogen, mineral oil, or water. At least three methods used for most strains.

Special services: 2000–3000 identifications (mainly pure cultures) undertaken by a team of 16 taxonomists. International Depositary for patent strains under the Budapest Treaty. Safe-deposit service. The Medical Mycology Department undertakes identifications and analysis of medical specimens. Contract research offered in various fields (food mycology, biodeterioration tests, mould problems in houses and working environments, development of bioinsecticides for control of insect pests).

Research: The emphasis is on the systematics of all groups of fungi, with multidisciplinary approaches (ultrastructure, serology, electrophoresis, physiology, GCMS, etc.). Applied research related to the biological control of pests, wood preservation, agriculture, plant pathology, and medical mycology.

Training: Three courses are offered: General Mycology (four weeks), Introduction to Food-borne Fungi (two weeks), and Medical Mycology (four weeks). Joint courses with IMI (see below) on Fungi and Food-spoilage and Biodeterioration (two days). Specific training available. Postgraduate training for MSc and PhD.

Publications: The collection produces a *Newsletter*, guides to courses, and the serial publication *Studies in Mycology*. Full list available on application.

Notes: The oldest service collection of fungi in the world. Has a collaborative agreement with the CAB International Mycological Institute (IMI).

Culture Collection of Basidiomycetes   CCBAS
Department of Experimental Mycology,
Institute of Microbiology
Czechoslovak Academy of Sciences, Videnska 1083
142 20 Prague 4, Czechoslovakia
Tel.: Prague 47-21-151 or 47-19-740
Founded: 1960   *Strains of fungi:* 500
Status: Sponsored by the Czechoslovak Academy of Sciences.
Coverage: Basidiomycotina only.
Culture sales: Cultures may be purchased, or in some cases exchanged for other cultures.
Research: The cultures are used for research purposes with

regard to their physiology and biochemical activity, and they also serve as comparative material for other studies.

Notes: Part of CCM (see below). The cultures were mainly isolated by the explant method from fruiting-bodies collected in Czechoslovakia. A small part of the collection was received from other institutions on an exchange basis.

Culture Collection of Fungi   CCF
Department of Botany
Charles University, Benatska 2, 128 04 Prague 2
Czechoslovakia
Tel.: 297941-9
Founded: 1964   *Strains of fungi:* 1200
Coverage: Ascomycotina, Deuteromycotina and Zygomycotina.
Catalogue: Catalogue available.
Special services: Carries out investigations for scientific and industrial institutes. Preserves ex-type and patent cultures.
Research: Concentrates on the taxonomy and ecology of members of the filamentous fungi (Mucorales, Moniliales, Ascomycotina).

Czechoslovak Collection of Microorganisms   CCM
J. E. Purkyne University, tf. Obrancu miru 10
662 43 Brno, Czechoslovakia
Tel.: 23407
Founded: 1963   *Strains of fungi:* 520
Status: A public collection.
Coverage: The majority of fungi held are Hyphomycetes, Coelomycetes and Ascomycotina, including several species of Oomycetes, Zygomycotina and Basidiomycotina.
Catalogue: Catalogue available.
Culture sales: Provides cultures for research institutions, schools and industry in Czechoslovakia as well as abroad, for a fee.
Preservation: The principle method of preservation is freeze-drying.
Special services: Cultures that are the subject of patent procedures are accepted for deposit.
Notes: Cultures held are of importance to plant pathology, human and veterinary medicine, industry, physiological and biochemical research, biodeterioration studies and as producers of various metabolites, enzymes, antibiotics, mycotoxins, etc.

Collection de Champignons et d'Actinomycètes Pathogènes de la CNCM   CNCM
Unite de Mycologie, Institut Pasteur
25 rue du Dr Roux, 75724 Paris Cedex 15
France
Tel.: (1) 45-68-83-57; Telex: 250.609 F

Strains of fungi: 1500

Coverage: Filamentous fungi and yeasts pathogenic to man, and a restricted number of common fungi that may be isolated with the pathogens as contaminants.

Catalogue: Regularly updated catalogue available.

Preservation: Mainly preserved by lyophilization and in liquid nitrogen.

Special services: An identification service is provided by the Unit of Mycology, but limited to fungi that are implicated in human pathology.

**Deutsche Sammlung von Microorganismen
(German Collection of Microorganisms)   DSM**
Mascheroder Weg 1b, D-3300 Braunschweig
Federal Republic of Germany
Tel.: 0531-6187-0; Telex: 952667 GEBIO D
(Electronic Mail: TELECOM GOLD 75: DBI0178)

Founded: 1968   *Strains of fungi:* 1400

Status: The DSM is the national service collection of micro-organisms, based in Göttingen until 1987.

Coverage: Main emphasis is on cultures relevant to biotechnology, education and applied microbiology. Strains of all groups are held.

Catalogue: A regularly updated catalogue provides information on each strain; computerisation of strain data will be started within the MINE network of which this is the German node (see Chapter 2).

Special services: The DSM offers identification, consultation, safe-deposit, and freeze-drying services and is an International Depositary Authority for strains cited in patent applications.

Research: Projects are conducted on problems related to the work of the collection and taxonomy.

CAB International Mycological Institute   IMI
Ferry Lane, Kew, Surrey TW9 3AF, UK
Tel.: 01-940-4086; Telecom Gold/Dialcom: 84: CAU009;

Telex: 265871 (MONREF G) or 847964 (COMAGG G) quoting CAU009

Founded: 1920 *Strains of fungi:* 12 500

Status: Formerly the Commonwealth Mycological Institute. Incorporating the UK National Collection of Fungus Cultures (from 1947) and the Biodeterioration Centre (from 1984). An Institute of CAB International; international legal status; sponsored by 29 governments; The Culture Collection and Industrial Services Division is also supported by UK Department of Trade and Industry. The UNESCO MIRCEN for Mycology. An Associated Institute of the University of Reading. See Hawksworth (1985*b*).

Coverage: All groups (except Hymenomycetes and human pathogens), especially isolates of industrial, agricultural, medical, forestry, biodeterioration, and systematic importance. Extensive data available on the physiological and biochemical attributes of selected strains.

Catalogue: Hard copy revised every 2–3 years. Catalogue (including strain data) computerised and available on-line to outside users (further details available on request). The UK Node for MINE and contributor to MiCIS (see Chapter 2).

Culture sales: World-wide. Reductions for non-commercial use in member countries. Biodeterioration test sets. Exchanges considered.

Preservation: Methods used include freeze-drying by centrifugal and shelf-driers (8000 strains), liquid nitrogen, mineral oil, silica gel, soil or water. At least two methods used for most strains.

Special services: 8000 identifications undertaken each year by a team of 18 taxonomists. The Biodeterioration and Industrial Laboratory undertakes mould growth testing to British and other standards, fuel testing, consultancy and advice; tropical testing facility; International Depositary for Patent Strains under the Budapest Treaty. Safe Deposit service. The Biochemical Laboratory undertakes screening for particular products or properties, testing for anti-microbial activity and microbial pesticidal properties (in collaboration with the Jodrell Laboratory, Royal Botanic Gardens, Kew). The Information Services provide on-line and manual searches on specific topics and also annotated bibliographies. A Plant Pathology Laboratory provides liaison and consultancy services, especially on tropical crops.

Research: Emphasis is on the systematics of filamentous fungi of applied importance (especially using biochemical and ultrastructural techniques), preservation methods and the microchemical determination of secondary metabolites.

Training: Courses range from one day to 6 weeks on a wide range of mycological topics. Specific training available. Postgraduate training to MSc and PhD.

Publications: In addition to research papers, the Institute produces a *Newsletter*, ten serial publications, and one to two books each year. Of special interest to biotechnologists are *Biodeterioration Abstracts* (quarterly), the *Review of Medical and Veterinary Mycology* (quarterly), Kirsop (1980), and Smith & Onions (1983). Full list available on application.

Notes: The IMI, with a staff of 70 (23 PhDs), is the largest centre concerned with systematic mycology. The culture collection is a part of the Culture Collection and Industrial Services Division of the Institute; other divisions are for Taxonomic and Identification Services and Information Services. The services are backed by over 300 000 dried reference specimens of microfungi, a library of over 6000 volumes and 150 000 reprints. The IMI has collaborative agreements with the Centraalbureau voor Schimmelcultures, Baarn (CBS), Biosystematics Research Centre, Agriculture, Canada (BRC), Institute of Seed Pathology in Developing Countries, Copenhagen, and the Aquatic Phycomycete Culture Collection, Reading, UK (see above).

IPO-Collection of Fungal Pathotypes   IPO
Research Institute for Plant Protection,
Binnenhaven 12, PO Box 9060, 6700 GW
Wageningen, The Netherlands
Tel.: (08370) 19151
Cable: IPO-Wageningen   *Strains of fungi:* 1317

Coverage: Strains and races of pathogenic fungi (inoculation material supplied to plant breeding institutes and commercial plant breeders) relevant to the following crops: barley, Brassicae, Cucurbitaceae, lettuce, onion, *Phaseolus* and *Vicia* beans, potato, spinach, tomato and wheat. Race collections are maintained of *Bremia lactucae, Colletotrichum lindemuthianum, Fulvia fulva, Fusarium oxysporum* f.sp. *conglutinans,* f.sp. *cucmerinum,* f.sp. *lycopersici,* and f.sp. *pisi, Peronospora farinosa* f.sp. *spinaciae, Puccinia striiformis,* and *P. recondita.* Also other species from Zygomycotina, Basidiomycotina and Deuteromycotina.

Catalogue: Catalogue available.

Notes: Collection maintained to serve resistance breeding.

Laboratoire de Souches de Cryptogamie du Museum National d'Histoire Naturelle   LCP
12 Rue de Buffon, 75005 Paris, France
Tel.: (1) 43-31-35-21 or (1) 43-31-95-11   *Strains of fungi:* 2500
Coverage: Conidial fungi and Mucorales, Ascomycotina, Deuteromycotina, Zygomycotina and Basidiomycotina. Most of the strains (soil fungi, biodeteriogenic fungi, antagonistic fungi, coprophilous fungi) have been and continue to be isolated by staff of the laboratory.
Catalogue: Catalogue available.
Preservation: All strains are preserved by culturing on to agar, but lyophilization and cryopreservation are also used when possible.
Special services: Some isolates are tested for metabolite production and other activities (biodeterioration, antagonism). The collection includes strains used for standards tests of the Association Francais de Normalization (ANOR).

Mycothèque de l'Université Catholique de Louvain   MUCL
Place Croix du Sud 3/8
B-1348 Louvain-la-Neuve, Belgium
Tel.: 32-10-43 37 42; Telex: UCL B 59037
Founded: 1894   *Strains of fungi:* 15 000
Status: Founded by the Catholic University of Louvain, and including collections from Biourge, Meyer and Simonart. Integrated in the Laboratory of Systematic and Applied Mycology, Faculty of Agronomy. Financed by the Belgium Science Policy Office in the framework of the Coordinated Collection of Microorganisms project.
Coverage: All groups of fungi, especially Hyphomycetes, including 200 species of yeasts and fungal isolates of industrial, agricultural, forestry, biochemical and biodeterioration importance, as well as of taxonomic interest.
Catalogue: Data for forthcoming catalogues are being computerised. The printed catalogue is planned to be published in 1987 and will be on-line via the MINE network.
Culture sales: World-wide. Reductions for non-profit institutions, educational institutions and developing countries. Exchanges considered.
Preservation: Methods used include freeze-drying, mineral oil, soil, and distilled water.

Special services: Identification is carried out by a team of eight taxonomists. Biodeterioration testing of materials, screening for the production of primary and secondary metabolites, physiological studies and fermentation studies. Industrial and legal expertise. Information service based on a specialised library. Exclusive deposit of strains for safe keeping by industrial users.

Research: Fundamental mycology: morphology, cultural study (growth, sporulation, pigmentation, etc.), taxonomy and nomenclature. Applied mycology: food preservation, fermented foods, composting, edible mushroom cultivation, cryopreservation of fungal cultures.

Training: University lectures in general and systematic mycology. Specific training in taxonomic research, applied mycology and culture maintenance.

Publications: Publication of *Mycotaxon*, an international journal on taxonomy and nomenclature of fungi and lichens; 2 volumes of 1200 pages published annually since 1974.

Friedrich-Schiller-Universität Jena   MW
Sektion Biologie
Pilzkulturensammlung, Freiherr-vom-Stein Allee 2
Postfach 16/329, Weimar 5300
German Democratic Republic
Tel.: Weimar 3498   *Strains of fungi:* 5400
Coverage: All groups except Mastigomycotina.
Catalogue: A catalogue is available.

National Collection of Pathogenic Fungi   NCPF
Mycological Reference Laboratory
Public Health Laboratory, 61 Colindale Avenue
London NW9 5HT, UK
Tel.: 01-200-4400; Telex: 8953942 (DEFENDG)
Founded: 1946   *Strains of fungi:* 600
Status: Part of the UK Public Health Laboratory Service.
Coverage: Fungi from infections of warm blooded animals, including dermatophytes and systemic pathogens (Category III).
Catalogue: A full catalogue is planned to be issued in 1987.
Preservation: Primarily by lyophilization and water storage; many also kept under mineral oil.

Training: Training in cultural techniques for use with pathogenic fungi can be given by arrangement.

Research: A taxonomic research programme involves the study of diverse groups of fungal pathogens.

National Collection of Wood Rotting Fungi   NCWRF-FPRL

Biodeterioration Section, Timber Protection Division
Building Research Establishment, Garston
Watford, Herts. WD2 7JR, UK
Tel.: 0923-674040; Telex: 923220
Founded: 1927   *Strains of fungi:* 550
Status: Sponsored by the UK Government, Department of the Environment. Transferred from Princes Risborough from 1 January 1988.

Coverage: Wood-rotting macrofungi, mostly basidiomycetes with a few larger ascomycetes, of world-wide origin.

Catalogue: Strains listed in own catalogue (charged), in MiCIS and MINE databases; species listed in the WFCC's World Data Center Directory.

Preservation: Cultures are maintained under liquid nitrogen, but are supplied on agar slopes.

Special services: Consultancy, advice on identification, contract work and biodeterioration testing specialising in wood, wood products and wood preservative biocides.

Portsmouth Polytechnic Culture Collection   PPCC

School of Biological Sciences
King Henry Building, King Henry I Street
Portsmouth, Hants. PO1 2DY, UK
Tel.: 0705-827681
Founded: 1971   *Strains of fungi:* 2000
Coverage: The culture collection is composed of aquatic fungi, primarily higher marine fungi. These fungi have been obtained from localities around the world. Some fruit under laboratory conditions and up to 10 strains of each species is held. Besides the marine fungi, some 500 freshwater Hyphomycetes are also held in the collection.

Catalogue: A catalogue is available.

All-union Collection of Microorganisms   VKM
Institute of Biochemistry and Physiology of Microorganisms,
USSR Academy of Sciences
Moscow Region, Pushchino, 142292, USSR
Tel.: 3-05-26 (Pushchino); 231-65-76 (Moscow)
Telex: 3205887, Oka
Founded: c. 1960   *Strains of fungi:* 2500
Coverage: All groups.
Catalogue: A catalogue is available.
Notes: Main interests are in general and industrial mycology.
Research, teaching and other services carried out.

1.5.5   *North America*

Atlantic Regional Laboratory   ARL
National Research Council
1411 Oxford Street, Halifax, Nova Scotia B3H 3Z1, Canada
Tel.: (902) 426-8070
Founded: 1956   *Strains of fungi:* 2000
Coverage: All groups.
Catalogue: Strain data are computerised and incorporated into
the Canadian Fungal Culture Collection Database (see Chapter 2).
Preservation: Cultures are maintained by continuous transfer,
storage under mineral oil, and by refrigeration at −15 °C.
Notes: Incorporates strains previously held at the National
Research Council's Saskatoon Laboratory. Almost 1000 of the strains are
known producers of biologically active metabolites.

American Type Culture Collection   ATCC
12301 Parklawn Drive
Rockville, Maryland 20852, USA
Tel.: (301) 881-2600
Telex: 898055 ATCC North or ATCCROVE 908768
(Electronic Mail: DIALCOM 42: CDT0004)
Founded: 1925   *Strains of fungi:* 22 000
Status: The American Type Culture Collection (ATCC) is a non-
profit organisation which houses the most diverse collection of organ-
isms in the world. It maintains over 40 000 different strains of animal
viruses, bacteria, bacteriophages, cell lines, fungi, plant viruses, protists
and yeasts, plus rapidly growing collections of oncogenes, recombinant

DNA vectors, hybridomas, plant tissue cultures, and human DNA probes and libraries.

Coverage: All groups of fungi. The principal functions include not only the accession, preservation, authentication and distribution of reference and ex-type cultures, but also the creation and manipulation of data files for each culture maintained.

Catalogues: Prepared individually for each of the ATCC collections, published regularly and available free of charge; postage is charged for shipments outside the USA. The catalogue of fungi and yeasts contains data on source, nomenclature, taxonomy, biotechnology, genetic information, availability, literature citations, culture conditions and permits required for shipping.

Preservation: In order to prevent loss of viability, contamination, variation, mutation or deterioration, all the cultures maintained in the ATCC are freeze-dried and stored at 4°C and/or frozen in liquid nitrogen.

Research: This includes comparative microbiology, microbial systematics, computer-assisted identification analyses, and improved methods for the isolation, propagation, characterisation and preservation of strains, strain improvements for industrial applications, eukaryotic gene regulation, studies on the mechanism of action of oncogenes and artificial cultivation of exotic mushrooms.

Special services: Culture identification, cryopreservation, freeze-drying of microorganisms and biological reagents, culture safekeeping, plasmid preparation, mycoplasma testing, karyotyping and virus purification. The ATCC was the first approved International Depositary Authority (IDA) under the Budapest Treaty on the International Recognition of the Deposit of Microorganisms for the Purposes of Patent Procedures, and is also an official depository for patent procedures under the European Patent Office (EPO), the United States, Germany, Japan, and many other countries.

Training: A workshop series offers scientists and technicians hands-on laboratory training in areas such as cryopreservation, quality control and assurance, recombinant DNA methodology, computer usage in microbiology and biotechnology patents.

Notes: The collection acts as an information clearing house in all disciplines of mycology and disseminates information on the availability of strains, fungal nomenclature, classification and characterisation, preservation techniques, and the application of fungus cultures to the study and solution of human problems involving food, energy, health and environmental pollution.

Canadian Collection of Fungus Cultures   CCFC
Biosystematics Research Centre, Saunders Building
Agriculture Canada, Ottawa K1A OC6, Canada
Tel.: (613) 996-1665
Founded: 1974   *Strains of fungi:* 12 000
Coverage: All major groups with significant numbers of moulds (Hyphomycetes), wood-rots (Hymenomycetes) and bird's-nest fungi (Nidulariales).

Catalogue: A catalogue is not available, but species are listed in the Directory of the WFCC's World Data Center.

Culture sales: Most isolates are available for distribution.

Special services: Identification, consulting, advisory and training services.

Notes: Important collections are the Nidulariales of H. J. Brodie, wood-rot fungi of M. K. Nobles, 'and Hyphomycetes of S. J. Hughes. The collection is part of the Canadian Fungal Culture Collection Database (see Chapter 2). It is supported by eight taxonomists and an extensive herbarium (DAOM).

Mycological Culture Collection   DLR
Mycology Laboratories
Wadsworth Center for Laboratories and Research
New York State Department of Health, Albany
New York 12201, USA
Tel.: (518) 474-2168   *Strains of fungi:* 450
Coverage: The collection primarily contains clinically important filamentous fungi. All groups are maintained except Basidiomycotina and Mastigomycotina.

Catalogue: No catalogue is available.

Culture sales: Subcultures of collection strains are provided free of charge to educational institutions, scientific investigators and non-profit organisations.

Preservation: Cultures are maintained under refrigeration and/or as lyophilized specimens.

Forintek Collection of Wood-inhabiting Fungi   EFPL
Biotechnology and Chemistry Department
Forintek Canada, Corp., Eastern Laboratory
800 Montreal Road, Ottawa, Ontario K1G 3Z5, Canada

Tel.: 613-744-0963; Telex: 053-3606
Founded: 1920   *Strains of fungi:* 2500
Coverage: Wood-inhabiting filamentous fungi.
Catalogue: A catalogue is available, cost deductible from the first purchase.

Notes: EFPL comprises the collections of the former Eastern and Western Canada Forest Products Laboratories. The traditional emphasis on wood-inhabiting fungi, primarily wood-decaying basidiomycetes, has recently expanded to include fungi with possible uses in the biological modification or protection of wood or wood products. Cultures are free to government or university research groups and for sale to others.

Fungal Genetics Stock Center   FGSC
Department of Microbiology
University of Kansas Medical Center
Kansas City, Kansas 66103, USA
Tel.: 913-588-7044
Founded: 1960   *Strains of fungi:* 5500
Status: Sponsored by the US Government (National Science Foundation).
Coverage: Ascomycotina.
Catalogue: Independent catalogue published (latest version in July 1986).
Culture sales: Cultures available for research and teaching.
Special services: Research and teaching.
Notes: Genetic stocks of *Aspergillus nidulans*, *Neurospora* species and *Sordaria fimicola* are maintained.

Agricultural Research Service Culture Collection   NRRL
Northern Regional Research Center
1815 N. University Street, Peoria
Illinois 61604, USA
Tel.: 309-685-4011
Founded: 1941   *Strains of fungi:* 44 000
Status: The ARS Culture Collection is the major microbial culture collection of the Agricultural Research Service, US Department of Agriculture. See Kurtzman (1986).
Coverage: The collection is comprised of strains primarily of

agricultural and industrial significance, but with relatively few human, animal and plant pathogens. The US Department of the Army Quartermaster (QM) collection is also maintained as part of the ARS Culture Collection.

Catalogue: No catalogue is issued. Requests may be made by asking for particular taxa, or by strain number if known, or for strains with particular uses.

Culture sales: No charges are made for cultures distributed from the general collection, but requests must be limited to 12 strains. A fee is charged for distribution of released patent cultures deposited after 1 November 1983. Earlier released deposits are sent without charge. Requests for cultures should be written.

Preservation: The majority of strains are maintained by lyophilized preparations. Cultures not amenable to lyophilization are preserved by freezing in liquid nitrogen.

Special services: The ARS Culture Collection is recognised as an International Depositary Authority under the Budapest Treaty. Identification of unknown strains is undertaken on a limited scale when compatible with the aims of the curators.

Research: Emphasis is on the systematics of economically important microorganisms through comparisons of molecular, biochemical and ultrastructural similarity.

Training: Specific training is available when compatible with research programmes of individual staff members.

Ontario Ministry of Health Fungal Collection   TPHL

Toronto Public Health Laboratories
Mycology Laboratory, Laboratory Services Branch
Box 9000, Terminal 'A', Toronto, Ontario M5W 1R5
Canada
Tel.: (416) 248-3330 or (416) 248-7478
Founded: 1946   *Strains of fungi:* 1100
Status: Wholly supported by the government of the Province of Ontario.

Coverage: All groups, except Mastigomycotina. Includes fungi pathogenic to humans and higher mammals (systemic, subcutaneous, superficial, and opportunistic pathogens), with particular emphasis on species occurring in Canada. Also includes saprobic fungi received as contaminants from clinical samples.

Notes: TPHL has particularly strong collections of *Blastomyces dermatitidis* and variants of *Trichophyton rubrum*.

University of Alberta Microfungus Collection and Herbaria UAMH
Devonian Botanic Gardens, University of Alberta
Edmonton, Alberta T6G 2El, Canada
Tel.: (403) 432-2311
Founded: 1960   *Strains of fungi:* 5200
Status: Located in Western Canada, the UAMH is the second largest collection of filamentous fungi in Canada. The UAMH serves as a world reference centre for the identification of unusual pathogens of medical importance.
Coverage: Fungi in the collection include all major taxonomic groups, but particular emphasis has been placed on mould fungi (Hyphomycetes). Fungi associated with human and animal disease are particularly well represented and the UAMH holds one of the major world collections of onygenalean fungi.
Catalogue: A catalogue is published. The latest issue was in 1986. Data is computerised.
Special services: Identification, culture depositary and distribution services, consulting and advisory services, training. UAMH is accessible to any scientist who needs to use the services of the collection or the collection itself. However, because of the small size of its staff, requests for service must be evaluated according to their relevance to the UAMH programme and the amount of time required. Requests for service should be addressed to the Curator.
Notes: Probably the largest collection of medically important fungi, especially dermatophytes and Gymnoascaceae.

Upjohn Culture Collection UCR
The Upjohn Company
301 Henrietta Street, Kalamazoo, Michigan 49001, USA
Tel.: (616) 385-7158/224401
Telex: TWX 810-2772602
Cable: UPJOHN
Founded: 1954   *Strains of fungi:* 3000
Status: Sponsored by industry.
Coverage: All groups. The objective of the collection is to maintain strains for assay, research, reference and production use.

Catalogue: An in-house catalogue is produced.

Culture sales: Cultures are distributed outside The Upjohn Company on a limited request basis. These are cultures cited in publications and/or patents.

Preservation: Cultures are maintained in sterile sandy loam at 4 °C, and as agar plugs in sterile straws placed in screw-cap vials. The vials are stored at −150 °C in the gas phase of liquid nitrogen.

Notes: Some cultures restricted. Genetic strains maintained.

1.5.6 South America

Departamento de Micologia, Fundaçao Instituto Osvaldo Cruz   OCF

Av. Brasil 4365, 21040 Rio de Janeiro, RJ, Brazil

Tel.: (21) 280-8787 R 303/241

Founded: 1922 *Strains of fungi:* 1700

Coverage: A research and service culture collection with 112 genera, especially of medical interest.

Catalogue: Holdings listed in the national catalogue 'Catalogo Nacional de Linhagers' (Fundaçao Tropical de Pesquisas e Tecnologia 'André Tosello', 1986).

(Electronic Mail: DIALCOM CDT0094)

Notes: Preservation is by subculturing.

1.6 Development of resource centres

It will be evident from the preceding sections that many countries still lack major culture collections of filamentous fungi. Steps to rectify this situation are taking place in many countries, which are assessing national needs, developing or reorganising existing collections or establishing new ones.

There is a substantial cost in providing the wide range of services culture collections can potentially offer to applied microbiologists. In order to provide extension services such as identification, preservation, consultancy, biological testing, screening, biochemical characterisation, training and patent deposition, skilled personnel are essential and resources to employ appropriate scientists are required. These revenue-generating operations need time to be advertised and become established and it may be several years before they operate profitably. In some cases national funding bodies find it beneficial to offer subsidised or free services in order to stimulate growth in microbiologically based industries.

The value of existing collections can be enhanced by development of an acquisitions policy. In order to do this, collections must decide whether to (a) specialise in material relevant to particular fields (e.g. industry, medicine, plant pathology, forestry), (b) meet the needs of educational establishments as well as research and industry, or (c) operate nationally or internationally. In these days of scarce resources and generally rapid mail services, national boundaries are decreasingly an obstacle, and it may be advantageous to see whether a country's needs can be met from existing neighbouring collections. Effort can then be put into building up complementary rather than carbon-copy sets of strains, thereby enriching the resource as a whole.

When developing microbiological resource centres, the lack of long-term preservation facilities may limit the groups of fungi that can be held, and additions to the range of techniques will extend the capability of individual collections. Again, the development of physiological and biochemical strain data is of major importance, but is rarely obtained, since identification of filamentous fungi seldom requires this kind of information (see Chapter 5). The amount of such data on strains held in culture collections, and thus available to the computer databases now being established, is often minimal. To carry out the necessary tests to obtain the data is a gigantic operation, particularly when it is considered that in the genus *Penicillium* alone, over 130 physiological and biochemical features give differential results (Bridge *et al.*, 1986).

Nevertheless, collections need to make catalogue and strain information available in both hard copy or machine readable form if it is to be readily accessed by applied microbiologists. Resources need to be provided to individual collections to enable them to carry out the necessary work and collaborate with national and international microbial database projects such as MiCIS, MINE and MSDN (see Chapter 2).

1.7 Safeguarding resource centres

The conservation of the genetic resources maintained by culture collections is of necessity a long-term operation, the objective being the preservation of living material for posterity. This is not always appreciated by funding bodies, and many collections have suffered as a result of a succession of short-term support grants, which have not permitted long-term development plans to be initiated or the resource developed to increase its value to users. The cost of maintaining cultures is usually a fraction of the total cost that would be incurred in reisolating the strains from nature. Secure, long-term funding is therefore essential to safeguard culture collections if past investment is not to be lost.

When funding has to cease, adequate provisions need to be made to enable important strains to be incorporated into another collection. Major service collections are often willing to take over endangered collections, but may require additional funding to do so if the collection to be absorbed is substantial or special maintenance procedures would have to be introduced. Research collections (see Section 1.4.3) are especially vulnerable, and when research is initiated that is likely to lead to the assemblage of important strains, provision for the deposit of these should be made in the initial research grant. Many isolates obtained at considerable public expense are lost to posterity when research grants cease.

An Endangered Collections' Committee has been established by the WFCC and may be approached by individuals or organisations which face difficulties with the continuing maintenance of collections. This Committee has resources from UNESCO, UNEP and IUMS to enable specialists to visit endangered collections, assist with documentation, help identify suitable recipient collections, and contribute to transfer costs and short-term emergency preservation. The Committee exists to respond to requests for assistance and will do all in its power to ensure the continued existence and availability of internationally important strains.

2

Information resources

M. I. KRICHEVSKY, B.-O. FABRICIUS and
H. SUGAWARA

2.1 Introduction

Microbiologists are faced with consideration of exponential growth in their laboratories on a daily basis. As users of a chapter on information resources for biotechnology they are exposed to a double dose of exponential growth. First, the explosion of information technology itself is due to the massive amounts of computing power available at ever diminishing cost. In turn, a population of computer-aware and computer-literate microbiologists present a growing demand for more sophisticated access to modern information technology. The community of information technologists in concert with microbiologists are responding to this demand with a multiplicity of initiatives using various strategies.

The resulting activity induces feelings of inadequacy in the authors of such chapters as this, since at the moment of delivery to the editors the information is out of date. Resources previously known only by rumour are tested. Simple facilities being tested as pilot projects are quickly made available to the community. Local data banks open their doors to regional and even world-wide participation. Databases on databases spring up because of the need to discover available resources. Occasionally, resources fall by the wayside. The net result is an ever increasing base of information resources for biotechnologists.

While the information about information presented in this chapter is out of date as soon as it is written, the resources described are most likely to be improved and be more useful than the descriptions indicate. For information on new developments the listed resources should be contacted.

2.2   Information needs

The need of the biotechnologist for widely disparate categories of information is a consequence of the varied nature of the tasks required to design, develop, and consummate a process. The biotechnologist must find or develop genotypes of the required composition, discover the conditions for expression of the desired phenotypic properties, maintain the clones in a stable form, and describe all of these parameters in a fashion understandable to peers. Most of the categories of information required will be outside the expertise of any one individual. Thus, a panel of experts representing all disciplines involved must be assembled or access to diverse databases must be achieved. The library or publicly accessible databases can lead to the required information sources which range from the traditional scholarly publications to assemblages of factual or primary laboratory observations. This chapter presents an overview of the kinds of information available and the mechanisms to access them. In particular, it will concentrate on information resources for finding material with the desired attributes, be they taxonomic, historical, genetic, or phenotypic.

2.2.1   Interdisciplinary information sources

The personal training and professional experience in the particular narrow field in which they work allows microbiologists to perform many daily tasks without reference to outside sources of information. However, the interdisciplinary nature of the practice of biotechnology forces the use of navigational aids to the knowledge base of unfamiliar areas. This point is demonstrated by the observation that approximately 80% of the enquiries to the CODATA/IUIS Hybridoma Data Bank (see below) are from persons who are not immunologists. It follows that a successful database resource should be designed with the interdisciplinary users in mind as they often will form the largest segment of the user population.

The main pathways to locating an existing source of strains with the properties that will be useful in the projected process are through records of the primary observations of properties or through derived information such as taxa or strain designations. In either case, the desired result is one or more strain designations and instructions on where to get cultures. Even though the desired end result is the same with both pathways, the mechanisms for recording and disseminating the information are usually, but not necessarily, quite different.

Culture collections which have a stated mission of providing service to a public user community, especially through distribution of cultures outside of their host institution, tend to use a taxonomic orientation in that their records are commonly kept as discrete strain descriptions, often one strain to a page. The whole strain description is easily read while comparisons among strains are difficult.

Culture collections serving predominantly as local institutional repositories of strains for research, teaching, or voucher specimens for archival storage, tend to rely directly on primary observation data kept in tabular form with the attribute designations as column headings and the strain designations as row labels. In contrast to the previous case, comparisons among strains are easily made, while assembling a complete strain description may require following the row designation for the strain through multiple tables.

Even now, most service collections use traditional paper-based data management methods rather than computers. Within a few years, the majority of collections will be using computers as their main data-handling tool. The functional distinctions between these alternative forms of data organisations blur with the use of computers, but can still be a factor if good information management practices are not followed.

2.2.2 *Where to get strains*

The ultimate source for strains with desired properties is isolation from nature. Indeed, large efforts have been mounted to find strains with desired properties such as production of antibiotics. These efforts are feasible when a good screening procedure, such as zones of clearing around a colony, is available. Even then, the effort is labour-intensive with resulting high costs.

The existence of culture collections with data on the characteristics of the holdings makes possible enormous savings when appropriate strains are available. However, the data must be available to the process developer with reasonable ease. The best source for strains will often be from the collection with the most available data rather than the most complete selection of strains with likely sets of characteristics.

2.2.3 *How to get strains*

The pathways used in obtaining access to the required data for finding desirable strains start with the same foci as the collections themselves. A taxonomic strategy or primary observation strategy may be used. A taxonomic search strategy for strains producing higher concentrations of a particular material (e.g. penicillin, riboflavin, ethanol) might well begin with asking service culture collections for all strains of *Penicillium notatum, Ashbya gossypii, Saccharomyces cerevisiae* in their collections and screening them for level of production. This method of searching requires the searcher to know which taxa are likely to have the desired attributes.

A primary observation search strategy for organisms with the ability to degrade a particular material might well start with asking for all strains that degraded that material and had the growth characteristics that were desirable under the projected process conditions. The question to the collection might be 'Could you provide me with the characteristics of all your strains able to use hexadecane while growing aerobically at 25°C?'. This method of searching requires no taxonomic knowledge on the part of the searcher. Clearly, the format of data storage in the collection would markedly affect the relative ease of answering these questions.

2.2.4   *Strain data*

The traditional classes of data used to describe strains in culture collections still predominate. These include morphological, physiological, biochemical, genetic, and historical data. Clearly, these classes will form the basis for all future collection data as well. The current emphasis on biotechnology places new demands on the informational spectrum desired of microbial and cell line collections. In addition to the requests for taxa with specific properties, information is also requested on potential utility of strains in biotechnological processes and the actual or potential hazards arising from their use. These ancillary but important data are sparse in their availability, but some service collections with a concern for the needs of industry are collecting data of attributes specifically useful in biotechnological processes such as temperature tolerances, behaviour in fermenters, stability of monoclonal antibodies under adverse conditions, toxicity of products, and pathogenicity for unusual hosts.

Physiology. The bulk of information included in publicly accessible databases will be physiological and biochemical. Most queries will be on what the strains may be capable of doing with respect to the desired process, and the databases will be heavily weighted towards containing this kind of information.

While printed compendia of physiological and biochemical data are technically possible, they are rare and, in view of the expense of publishing such volumes, will continue to be rare. Rather, these kinds of data are more reasonably compiled in bulk and disseminated through computers.

Morphology. The detailed morphological data held by most culture collections are generally of less interest than details of physiological data for process development. In various broad classes of organisms (fungi, protozoa, and algae) morphology may be critically important for tax-onomy and identification. These same detailed attributes usually have little bearing on the conduct of most processes. However, some basic morphological information will be of fundamental importance in pro-cess engineering. For example, cell size knowledge is needed if filtration is a part of the process. Likewise, use of filamentous strains will require more energy for stirring fermenters than non-filamentous forms. Addi-tionally, knowledge of sexual reproduction or fusion attributes are critical if hybridization or strain improvement programmes are to be carried out.

Publicly accessible electronic databases used in searches will, with few exceptions, have only basic or general morphological information on strains because of the costs associated with printing or storage of such information. The decision on how much morphological detail to include in a public database will vary with the interaction between cost and importance of the information.

Industrial. Data on the use of particular strains for specific processes is a frequently requested category of data. Such data exist, but in a widely scattered, uncoordinated fashion. They are contained in the open literature, collection catalogues, patent disclosures, and other more obscure repositories. The same is true of related data on strain behaviour in the processes themselves. There are few biological data compendia equivalent to the materials properties databases available to the engineers in chemistry, metallurgy, and similar disciplines. This situa-tion is partly due to the nature of biological material and partly due to the later development of high technology in biology than in the other disciplines, in spite of the ancient history of biotechnology.

Hazard. Risk assessment of biotechnological activities concerns govern-mental regulatory bodies throughout the world. All of the concerns are environmental and may be in such forms as a specific disease of humans, animals, or plants, an undesirable imbalance in the environ-ment, or production of an undesirable product.

The data available to answer queries in the area of risk assessment are

quite sparse. The most common category of useful data is the pathogenicity of strains. This is largely a strain-specific phenomenon, the degree
of virulence varying from strain to strain, especially on serial propagation. Less commonly available are data describing strain persistence in
various environments and toxicity of products and rarest of all perhaps
are data to predict the effect of the introduction of strains into new
environments.

2.2.5   *Taxonomic data*

The traditional and important method of constructing databases
in service culture collections is with taxonomic orientation. The storage
of the data and their public presentation considers the strains first as
representatives of their taxa. Further, the data given in the description of
each strain in the catalogue of collection holdings are sparse as they
depend on the assumption that the reader knows, or is adept at finding,
the usual attributes of the taxon. Many clinical microbiology laboratories
only save the phenotypic data on antibiotic resistance patterns and the
putative name of the isolates; the data used to decide the name are
discarded. The records may indicate that the isolate is 'atypical' without
any indication of the attributes that led to this description.

Given the traditional organisation of culture collection catalogues, the
most common questions asked of service culture collections are likely to
be on specific attributes of strains listed in the catalogues, or the
companion question on the availability of strains with specific
attributes. Indeed, experience shows that service collection personnel
quickly develop great skill in searching the collection database by any
and all means available to answer such questions from the community.
In turn, the community learns that the collections are a valuable source
of various kinds of information beyond that contained in the catalogues.

The taxonomic orientation is natural in managing diverse culture
collections and determining the boundaries of interest for many spe-
cialty collections. Within a large collection, curator responsibilities are
usually assigned along taxonomic lines.

Historically, many service collections were established for studying or
supporting the study of taxonomy. This essential function still underlies
a great deal of service collection activity.

Many of the laws and regulations concerning shipment, hazard,
standards, laboratory safety, patents, use of biological material are
stated all or in large part in taxonomic terms. The US Federal Register
specifies strain and species in defining which strains are to be used as

standards for antibiotic testing. The attributes of these strains are not listed. Similar regulatory documents exist for other parts of the world.

2.2.6   *Regulations*

The development of regulations for efficient and safe use of biotechnological processes to the common good is heavily dependent on adequate data of diverse types. Unfortunately, databases do not exist to allow detailed appraisal of the potential hazard on a case-by-case basis. The use of taxonomic levels to evaluate and regulate is contraindicated by the very nature of the process of establishing taxa. The only solution to this dilemma requires the gathering and dissemination of appropriate data.

2.3   Information resources

An informal infrastructure of information gatherers, managers, and disseminators exists to answer questions on the practice and regulation of biotechnology as it relates to microbial strains and cell lines. It forms a very useful resource in spite of its informality. Further, a number of initiatives are under way which aim to manage this support system for biotechnology in a more formal and complete fashion.

The infrastructure exists independently of the practice of biotechnology since the same needs for information transfer are basic to all of microbiology and cell biology. Collection holdings are raw material for these sciences and new information initiatives are stemming from the application of molecular biology and advances in biotechnology. These developments are merely refining the information pathways that evolved before modern genetic engineering focused the public eye on one of the oldest forms of manufacturing, the use of biological materials in processes.

The historical sequence of development of the infrastructure started with the culture collections, proceeded with catalogue production, followed by collections of strain data in computers and, most recently, the creation of national, regional, and finally international, data services.

The ultimate source of data needed by the biotechnologist is the laboratory records of the collections. All other elements of infrastructure function to make these data accessible. The entry points to the pathways at all levels are the same as those considered previously: taxonomic or by detailed pattern of attributes. A combination is possible. The question 'Do you have a pseudomonad that degrades hexadecane?' will eliminate all yeasts that have the same ability.

2.3.1   Culture collection catalogues

Service collections publish catalogues describing their available strains to inform the public of their holdings and the salient properties of strains. A secondary effect is to reduce some of the labour overhead involved in answering questions from the public. A number of nations (such as Japan, China, and Brazil) have prepared combined catalogues, eliminating the need to obtain and consult multiple catalogues.

By the very nature of printed catalogues they function imperfectly since the information describing each strain is limited. Only one or at best a few attributes can be indexed so that detailed searching is impossible. Because of the positive accessioning policies of the service collections, catalogues are out of date upon publication. In general, the taxonomic entry route is served well, but the attribute route is necessarily left to follow-up questions to the collection itself. The most important deficiency is that only a very small proportion of the world's collections publish catalogues at all, since provision of a public service is not their prime function.

Where catalogues do exist, they form an important resource for the biotechnologist. The information they contain is carefully presented. If a taxon is known, much time can be saved by consulting a catalogue for availability. Often, valuable ancillary information on use, literature references, propagation conditions, or patents is included in the catalogue. Finally, the staff of the collection listed in the catalogue can be contacted for further information on their holdings or as entry into the rest of the informal information services of the collection community.

2.3.2   Individual collections

Culture collections of importance in biotechnology are not limited to the recognised service collections. In many cases, the biotechnologist must have access to a detailed collection of strains with the final selection of the particular strain for use in the process decided by personal comparative testing. Service collections may serve in this respect, but because of their usually broad nature cannot always maintain the specified holdings of the personal research or survey collection.

The mechanics of obtaining information from primary records of individual collections may not be simple. The curator must scan tables or individual strain records to match the pattern of the query. The process is faster for those collections which keep their records in a computer. While computer-aided searching is faster, considerable time still is required of collection personnel to search on a query by query basis.

Time spent in searching the database to answer public queries may be considered a normal and reasonable function by the service collections or a burdensome chore by collections with other basic missions. Either way, such searches can represent a considerable overhead on staff time.

2.3.3 Strain data compendia

The availability of computers for both record-keeping and data analysis has resulted in compendia of strain data in locations other than the collections. Hospitals have contributed antibiotic resistance data to large-scale surveys of changes of resistance plasmid distributions in bacterial populations. Numerical taxonomists assemble large databases of strain data in the course of their activities, often containing data contributed by other research workers. Ecologists and public regulatory agencies conducting surveys of the environment, including habitats such as soils, waters, foodstuffs, and wild and cultivated plants, often amass considerable amounts of data which are installed in some computer resource for management and analysis. The result is that the data describing the strains reside in a different location from the strains.

Arising from all this activity is a body of curators of data in support of the curators of strains. The relevant information specialists may be associated with a collection and, where such data management exists, the biotechnologist's search is immensely facilitated. Since the taxonomic designation is managed in the computer in the same way as any attribute of the strain, the entry into the database can be taxonomic or by attribute pattern with equal facility. The problem of finding the strains of interest is largely reduced to the problem of finding the databases themselves.

2.3.4 National, regional, and international data resources

Scientists and technologists band together in organisations focused on their disciplines in order to exchange information. Many of these scientific and technical societies also become providers of services to their members and the public community. Some resemble guilds or unions in providing advocacy for improving conditions for their members. Recently, societies have come full circle in that they are providing, directly or through advocacy, informational resources in the form of publicly available databases.

Since most societies are national, the earliest of these database efforts were national as well. In some disciplines, national efforts were deemed so valuable that they became international in use. The Chemical

Abstracts Service of the American Chemical Society is a well known example. Geopolitical regions have their counterparts in scientific and technical activities. The most notable of these regional efforts is within the European Economic Community with a ripple effect to other countries in Europe.

Either by combining regional resources or by direct international efforts, world-wide database resources are being established in many disciplines. Some stand alone and others are distributed in networks.

Informational resources describing culture collections and their holdings are distributed through all three levels as are the organisations concerned with sponsoring such resources.

National and regional organisations concerning culture collections. Microbiologists having interests in culture collections have banded together in national federations for culture collections. In some cases, cell biologists also are included. The boundaries for membership seem generally loose. The countries that have such national federations are listed and further information is provided in Chapter 8. The countries are: Australia, Brazil, Canada, China, Czechoslovakia, Japan, Korea, New Zealand, Turkey, United Kingdom, and United States of America. Japan, Brazil, and the UK are most actively pursuing national databases on collection holdings. Others are being discussed or planned. The Japanese database is designed to contain information from all types of microbial collections. The Brazilian and British systems are initially designed to concentrate on service collections. The systems limited to service collections could be expanded to include other collections as resources become available. More details on the data systems in the context of national and regional systems are given below.

Brazil

The second edition of the *Catalogo Nacional De Linhagens* was produced by the Fundação Tropical de Pesquisas e Tecnologia 'André Tosello' in Campinas, Brazil. The species names and designations of strains held by 23 Brazilian collections are listed in the catalogue. The catalogue includes bacteria, filamentous fungi, yeasts, protozoa, algae, animal cell lines, viruses, and miscellaneous unidentified microorganisms.

The database of the catalogue is being installed on the Brazilian national information system Embratel. Thus, it will be accessible through international telecommunication systems.

For more information contact:
Fundação Tropical de Pesquisas e Tecnologia 'André Tosello'
Rua Latino Coelho, 1301
13.100 – Campinas – SP
Brasil
(Electronic mail DIALCOM 42: CDT0094)

Canada

Curators of eight major culture collections holding fungi in Canada have contributed data to a common system maintained at the Atlantic Regional Laboratory of the National Research Council. Data on 19 500 strains are included in the following categories: name, accession number, substrate from which isolated, year of isolation, maintenance, literature references to metabolite production, pathogenicity, availability, and source. Collaborating collections include the Canadian Collection of Fungus Cultures (CCFC), Forintek Collection of Wood-Inhabiting Fungi (EFPL), University of Alberta Microfungus Collection and Herbarium (UAMH), and the National Research Council. Availability of data is currently restricted, but general on-line access is envisaged for late 1987.

For further information, contact:
Atlantic Regional Laboratory
National Research Council
1411 Oxford Street
Halifax
Nova Scotia B3H 3Z1
Canada

European Culture Collections' Organisation

The European Culture Collections' Organisation (ECCO) is an active group comprised of major service collections in countries that have microbiological societies affiliated with the Federation of the European Microbiological Societies (FEMS). All the ECCO collections are also affiliated with the World Federation for Culture Collections. Thirty-three collections are members from Belgium, Bulgaria, Czechoslovakia, Finland, France, Federal Republic of Germany, German Democratic Republic, Hungary, Netherlands, Norway, Poland, Portugal, Spain, Sweden, Switzerland, Turkey, United Kingdom, and the United Soviet Socialist Republic.

ECCO was established in 1982 to collaborate and trade ideas on all aspects of culture collection work. Since service collections inherently are organised repositories of information as well as cultures, ECCO is a valuable resource for finding information of interest in biotechnology. Each of the collections produces a catalogue of holdings. Such catalogues often hold information beyond the listing and description of the strains held. Also, the curators are likely contacts for other, non-service, collections in their countries.

As intercollection communication pathways grow and become formalised within and between ECCO countries, the prospects for an electronic communication network encompassing all these countries grows as well. Such a network is being actively discussed among its members at this time.

> *For further information contact:*
> European Culture Collections' Organisation
> Czechoslovak Collection of Microorganisms
> 662 43 Brno
> tr. Obrancu miru 10
> Czechoslovakia
> (Electronic mail TELECOM GOLD 75: DBI0154)

Japan Federation for Culture Collections

Since 1951 the Japan Federation for Culture Collections (JFCC) has promoted cooperation among culture collections and individuals in Japan as well as internationally.

In 1953, the JFCC started collecting data on the holdings in Japan and completed the catalogue which listed about 22 000 strains from 144 research institutions. The strains in the list were reidentified by a project team organised by Professor Kin'ichiro Sakaguchi of the University of Tokyo, resulting in the publication of a series of JFCC catalogues in 1962, 1966, and 1968. Further publications are planned.

> *For further information contact:*
> JFCC
> NODAI Research Institute Culture Collection
> Tokyo University of Agriculture
> 1-1-1 Sakuragaoka, Setagaya-Ku
> Tokyo 156
> Japan

Microbial Resource Centres (MIRCENs)
Information on Microbial Research Centres (MIRCENs) will be found in Chapter 8. These centres are found in both developed and developing countries. Each has its special focus of interest, for which it acts as a regional centre.

National and regional data resources
Institute for Physical and Chemical Research
In Japan, the Institute for Physical and Chemical Research (RIKEN) carries out international activities, such as being the host institution for the World Data Centre for Microorganisms and a node of the Hybridoma Data Bank. These activities are carried out by the Life Science Research Information Section (LSRIS) and Japan Collection of Microorganisms (JCM).

On a national level, LSRIS has developed the National Information System of Laboratory Organisms (NISLO), a directory of Japanese collections and their holdings. The NISLO covers microorganisms, animals, and plants. In the case of microorganisms, the LSRIS closely cooperates with the JCM.

The following information and services are currently available from LSRIS:
(1) the number of laboratory animals used in Japan and their scientific names;
(2) microorganisms maintained in the member collections of JFCC;
(3) algae maintained in culture collections in the world;
(4) identification of deciduous trees;
(5) fundamental references for cell lines widely used in Japan;
(6) bibliographical information for plant tissue and cell cultures.

For further information contact:
LSRIS
RIKEN
2-1 Hirosawa, Wako
Saitama 351-01
Japan
(Electronic mail DIALCOM 42: CDT0007)

Microbial Culture Information Service (MiCIS)
One of the first (along with Japan) national data efforts is that of the Microbial Culture Information Service (MiCIS) in the UK. This initiative is orientated towards providing public access to primary observation

data while retaining the taxonomic orientation for computer entry of the data. The initiative is a cooperative effort between the UK Department of Trade and Industry (DTI) and the UK Federation for Culture Collections. The following description is quoted from a joint statement of purpose between MiCIS and MINE (see below).

> This database has been developed by the Laboratory of the Government Chemist (LGC) on behalf of DTI following consultation with industry. MiCIS initially contains all data currently available on strains including catalogue information, hazards, morphology, enzymes, culture conditions, maintenance requirements, industrial properties, metabolites, sensitivities and tolerances. Further categories of information will be added as the system develops. MiCIS will contain data from all UK national culture collections and discussions are currently taking place to include data from private and other European collections.
>
> . . . Subscribers will be able to use MiCIS on-line or via a postal or telephone enquiry service and will pay an annual subscription plus a charge related to use. The system allows subscribers to search in confidence for the source of a named organism, the known properties of a named organism and for unknown organisms displaying particular properties. MiCIS News, a quarterly newsletter reporting MiCIS and culture collection activities, is supplied free to all subscribers.

Further information on MiCIS may be obtained from:
Microbial Culture Information Service (MiCIS)
Laboratory of the Government Chemist
Cornwall House
Waterloo Road
London SE1 8XY
UK
(Electronic mail TELECOM GOLD 75: DBI0015)

Microbial Information Network Europe (MINE)
Within the European Economic Community (EEC) the regional activities are primarily focused within the programmes of the Commission of the European Community (CEC). They fund a number of initiatives, either as sole efforts or collaboratively when the initiative has a scope beyond

the confines of the EEC. One such initiative within the EEC is the Microbial Information Network Europe (MINE). MINE has the traditional taxonomic orientation of the service collections. The UK node has provided the following description of MINE in a joint (with MiCIS) statement of purpose.

> This EEC database is a computerised integrated catalogue of culture collection holdings in Europe and is prepared as a part of the EEC Biotechnology Action Programme. CMI [CAB International Mycological Institute], in collaboration with CABI [CAB International] Systems Group, is to act as the UK node, in parallel with national nodes being developed in The Netherlands, Germany, and Belgium; discussions with Portugal and France are currently under way.
>
> . . . it will not include full strain data but only a minimum data set.
>
> Once enquirers locate a culture, they will then be referred to the collection concerned, or to national strain data centres . . . for any detailed strain information if this is needed.
>
> Initially, enquirers will contact MINE nodes by mail, telex, or telephone. At a later stage, online services may also be available on subscription.

Further information on MINE may be obtained from any of the national nodes (Chapter 1.5.4) or:
Direction Biology, Radiation Protection, Medical Research
SDME 02-41
(DG XII/F/2)
Commission of the European Community
200 rue de la Loi
B-1049 Bruxelles
Belgium
(Electronic mail TELECOM GOLD 75: DBI0004)

Nordic Register of Microbiological Culture Collections
In 1984, the Nordic Council of Ministers initiated support for a Nordic Register of Culture Collections encompassing Denmark, Finland, Iceland, Norway and Sweden. The Register's scope is inclusive of all sizes and functions of collections from small personal collections to large service collections. The first three years of development is focused on

strains of importance to agriculture, forestry, and horticulture. The building of the microcomputer-based database is an undertaking of the Department of Microbiology of the University of Helsinki, Finland. Development of software has been carried out in co-operation with the Nordic Gene Bank for Agricultural and Horticultural Plants in Alnarp, Sweden.

> *For more information, contact:*
> NORCC
> Department of Microbiology
> University of Helsinki
> SF-00710 Helsinki
> Finland
> (Electronic mail DIALCOM 42: CDT0069)

International resources

The primary focus for international culture collection information resources is through the International Council of Scientific Unions (ICSU) with headquarters in Paris, France. Various components of ICSU have current or potential initiatives relating to providing information of interest to biotechnologists. These include: Committee on Data for Science and Technology (CODATA), World Federation for Culture Collections (WFCC), International Union of Immunological Societies (IUIS), and International Union of Microbiological Societies (IUMS).

Committee on Data for Science and Technology (CODATA)

The Committee on Data for Science and Technology (CODATA) was established in 1966 by the International Council of Scientific Unions (ICSU) to promote and encourage the production and international distribution of scientific and technological data. Its initial emphasis was in physics and chemistry, but its scope has been broadened to data from the geo- and bio-sciences. CODATA is 'especially concerned with data of interdisciplinary significance and with projects that promote international cooperation in the compilation and dissemination of scientific data'.

The main activities of CODATA are carried out by Task Groups established for specific projects. Of special interest to biotechnologists are the biologically oriented Task Groups on the Hybridoma Data Bank, Microbial Strain Data Network, and Coordination of Protein Sequence Data Banks. The first two are the policy boards of the activities while the last is a co-ordinating body among the existing sequence data banks.

For more information on CODATA contact:
CODATA Secretariat
51 Boulevard de Montmorency
75016 Paris
France
(Electronic mail TELECOM GOLD 75: DBI0010)

Hybridoma Data Bank (HDB)

In 1984, the CODATA/IUIS Hybridoma Data Bank (HDB) started building an international database on hybridomas, other immunoreactive cell lines, and monoclonal antibodies. The HDB is designed to act as a locator service, help avoid duplication of effort, and provide a research tool on relationships among reactivity patterns. An international infrastructure (with data bank branches in the USA, France, and Japan) is in place, a growing database is being assembled, and queries are being answered. A pilot project for a publicly accessible on-line service is in progress.

The most common queries involve the antigen–antibody reactivity/non-reactivity patterns. The possible antigens to be named cover all of biology, biochemistry, and a large part of organic chemistry. The complexities compound rapidly when such problems as tumour epitopes are at issue. The host taxonomy, organ, tissue, cell structure, developmental stage, pathology, antigen, and epitope all have nomenclatural ambiguities to some degree. Various authority sources, selected by exercise of best judgement, used in building a controlled vocabulary/glossary reduce the problems to manageable proportions. Communication paths, conventions, and frequent quality control exercises facilitate consistency of response.

The information on each cell line and product is quite comprehensive. In addition to reactivity attributes, considerable information is coded, such as fusion partners' histories, developer, availability, distributor(s), applications, assay procedures, antibody classes, immunisation techniques, literature and patent citations. Many of these categories have internal sub-categories.

Queries to the HDB may be through mail, telephone, or through the CODATA Network available on Dialcom which has nodes in 20 countries and links to the most common packet switching services as well as to the Telex and TWX systems.

On a pilot basis, a subset of the main database is available directly to the international public through the CODATA Network. The pilot subset emphasises antibody reagents readily available (commercially or otherwise) and their reactivities. Monitoring usage of this database yields a direct evaluation of the elements in the controlled vocabulary.

More information on the HDB is available from any of the three nodes:
Hybridoma Data Bank
12301 Parklawn Drive
Rockville, MD 20852
USA
(Electronic mail DIALCOM 42: CDT0004)

LSRIS
RIKEN
2-1 Hirosawa, Wako
Saitama 351-01
Japan
(Electronic mail DIALCOM 42: CDT0007)

CERDIC
Lab. d'Immunologie
Fac. de Médécine
Av. de Vallombruse
06034 Nice
France
(Electronic mail TELECOM GOLD 75: DBI0014)

Microbial Strain Data Network (MSDN)
The CODATA/WFCC/IUMS Microbial Strain Data Network (MSDN) was started in 1985 to construct a world-wide network of holders of strain data serving as Nodes in an informational network. The MSDN will act as a locator service for strains of microbes or cultured cell lines having specific attributes.

From the UNEP-initiated round table in 1982 (at the International Congress for Microbiology) where the concept was introduced, to the present time, the Network has been designed; an infrastructure has been established, consisting of a policy-making Task Group, six operating Committees, and a Secretariat; and pilot on-line databases have been installed on a publicly accessible commercial computer host. Oversight is by three components of the International Council of Scientific Unions: CODATA, WFCC, and IUMS.

The sheer magnitude of the number of repositories of microbial strain data and the numbers of strains held within those repositories generate serious data acquisition and communication problems. Collecting the desired data in one or a few places for general availability is a practical impossibility. Rather, the MSDN is designed to operate as a locator service for repositories of strains with desired combinations of attributes by an indirect method. Thus, the data repositories become Network Informational Nodes.

The Central Directory of the MSDN contains a list of the data elements coded by all the various Nodes rather than the data themselves. This database is a controlled vocabulary of standardised nomenclature of mainly biochemical and morphological features used for strain characterisation throughout the world. The initial basis for the controlled vocabulary comes from the CODATA-sponsored publication by Rogosa, Krichevsky & Colwell (1986). The user scans the vocabulary to select features of interest. The features in the form of the controlled terms are entered as search criteria into a second database which yields the names of repositories assessing the strains for possession of the desired features. When a Node is located which codes information on the desired data elements (features), the person querying the MSDN contacts the Node(s) directly for existing strains fitting the detailed pattern of attributes. The contact may be accomplished through telecommunications (where the capabilities exist) or by mail or telephone. In an increasing number of cases, direct access to host computer databases is possible.

Queries to the MSDN may be through mail, telephone, or through the CODATA Network available on Dialcom which has nodes in 20 countries and links to the most common packet switching services as well as to the Telex and TWX systems.

For further information on the MSDN contact:
Institute of Biotechnology
Cambridge University
307 Huntingdon Road
Cambridge CB3 0JX
UK
(Electronic mail TELECOM GOLD 75: DBI0001/DBI0005)

World Data Center on Microorganisms (WDC)
In 1984, the World Federation for Culture Collections (WFCC) officially reaffirmed the WDC as a component of the WFCC and thus accepted responsibility for the operation and management of the WDC. Because of the announcement of the retirement of the director of the WDC at

Brisbane, Australia, a public search was conducted for potential hosts to ensure continuity of the WDC effort. After competitive evaluation of the proposals, the Executive Board of the WFCC agreed that the WDC be transferred to the Institute of Physical and Chemical Research (RIKEN), Saitama, Japan.

WDC is an information centre which supports culture collections and their users. The primary tasks are:

(1) publication of descriptions of culture collections;
(2) production of a species-orientated directory of culture collection holdings.

The tasks of WDC are not limited to the above; other tasks will be performed based on necessity and available resources.

Information sources for WDC are culture collections and national/local/international data centres as well. The WDC will, for example, make use of, and cooperate with, HDB and MSDN (as it becomes operational).

For information dissemination, the WDC will use a variety of communication media such as mail, telex, cable, facsimile, electric mail, publication and magnetic devices (floppy disks and magnetic tapes).

The WDC currently holds information on 327 culture collections distributed over 56 countries. The core data of the WDC are descriptions of culture collections and their holdings, allowing users to locate a culture collection and/or taxonomic category of microorganism. They can find strains of specific species by consulting the list of species preserved in culture collections. The WDC database currently includes bacteria, fungi, yeasts, algae, protozoa, lichens, animal cells, and viruses.

The WDC is able to answer queries by any communication media except on-line retrieval. The WDC will provide on-line information retrieval in the future, depending on available resources.

For further information on the WDC, contact:
WDC/RIKEN
2-1 Hirosawa, Wako
Saitama 351-01
Japan
(Electronic mail DIALCOM 42: CDT0007)

World Federation for Culture Collections (WFCC)
The World Federation for Culture Collections (WFCC) provides infor-

mation to biotechnologists in a variety of ways within the overall mission of the WFCC as described in Chapter 8. Specific information initiatives described above are the World Data Center on Microorganisms and the Microbial Strain Data Network. Specialist committees established by the WFCC can provide information on a number of topics. The Patents Committee considers patent conventions. The Postal and Quarantine Committee is a good source of information on regulations governing shipment of cultures. The Publicity Committee publishes a newsletter on a periodic basis. The Committee on Endangered Collections identifies and takes action to rescue jeopardised collections. The Education Committee initiates training activities (e.g. books, videotapes, courses, individual instruction).

For more information on the activities and information resources of the WFCC contact the WDC (see above), the MSDN (see above), or any of the resource centres listed in Chapter 1.

2.4   Access to data resources
2.4.1   *Traditional*

Traditional methods for accessing data resources are still the most common and are quite satisfactory as long as the answers are not voluminous and are not needed quickly. Asking a culture collection curator about the availability of a single strain exemplifying a particular taxon or having a particular set of attributes will usually get a prompt reply. The query may be verbal, in person or on the telephone, or by mail. As the queries become more complex, the use of these methods become less satisfactory to both the seekers and providers of information.

The seeker of information becomes frustrated at the delay and incomplete nature of the answer that comes back. Often multiple cycles are required to refine the question to the point where the desired answer is given. While this refinement of communication is valuable in clarifying the true nature of the query (not always recognised from the initial inquiry) and common to all pathways, the frustration is amplified by the length of the cycle time.

The provider of information must devote increasing resources to answering queries. This takes professional expertise that could be used in the other work of the collection. Therefore, any mechanism which minimises the labour involved in answering queries increases the professional resources of the collection.

Both internal and external mechanisms are useful to alleviate some of the workload. Internal mechanisms include publications, such as the aforementioned catalogues, and computer management of the data for ease of searching and reporting. External mechanisms are primarily electronic forms of communication and are discussed in the next section.

2.4.2   Electronic

Except for voice communication via telephone, telecommunication has only recently become a part of collection life. Many of the larger collections have been using printed electronic messages (cables, Telex) for a while. However, the advent of computer-operated message transfer systems at reasonable cost (e.g. direct on-line access, electronic mail systems, and public packet switching services) have allowed economically practical electronic communication between questioner and answering resource.

Two parallel paths of development are taking place at this time in providing public access to data in collections. Some collections are following both paths simultaneously.

In the first instance, a collection may make its data accessible to the public by establishing access to a computer system maintained by the collection or its parent institution. Some examples are the Human Gene Probe Bank at the American Type Culture Collection, the National Collection of Yeast Cultures at Norwich, UK, Japan Collection of Microorganisms, Tokyo, and the CAB International Mycological Institute, Kew, UK.

The second approach is to install the data in a computer operated by others. The data and their installation may be accomplished under the control of the collection as is being done by the ATCC on the CODATA Network. Alternatively, all or part of the data may be installed on a computer operated as a national or regional facility such as those described above for Europe, Brazil, and Japan (MiCIS, MINE, Catalogo Nacional De Linhagens, and NISLO).

In all these cases of providing access through electronic computer services, the source of the data does not maintain the communication paths beyond the host computer. It is generally the responsibility of the seeker of information to find the most appropriate path. Unless the provider and seeker of information are at the same institution, where direct connection to the computer may be possible, the ordinary telephone system is likely to be the first resource connected to the seeker's terminal. Where short distances are involved, it may be reasonable to use only the voice-carrying telephone system. However, national and international data communications are using 'packet switching service' (PSS) for an ever-increasing share of data telecommunications. Such PSS transmission is cheaper and more reliable by far than direct telephone calls.

To find out more about telecommunications with the systems described in this chapter, get in touch with the appropriate system listed.

Help with establishing electronic communication paths may be obtained from the MSDN (see above).

3

Administration and safety

D. L. HAWKSWORTH and K. ALLNER

3.1 Supply of cultures

3.1.1 *Terms and conditions*

Service collections supply cultures from their collections with no terms or conditions of use attached. The purchaser is therefore free to use the culture in any way he wishes. These 'open' collections comprise cultures listed in catalogues that are available for sale to the public at stated prices.

However, not all cultures held in collections are available without restriction. In addition to patent strains, where special conditions apply (Chapter 6), or safe-deposit strains held on a confidential basis for a depositor (Chapter 7), strains isolated by collection staff or on which research has been carried out may also be held in 'reserve'. Such 'reserve' strains, which may be of potential industrial importance, are not included in catalogues nor made generally available to the public, but may be supplied on special terms on a case-by-case basis. They are usually supplied exclusively to a single industrial purchaser, together with data on their attributes. In addition to an initial payment for the exclusive rights, involvement in the further development of the strain and access to data derived, a small royalty on any resultant income may be stipulated. Some culture collections that have used their own resources to discover new biochemical or microbial activities of strains, feel it is appropriate that they receive some return should these strains prove to be commercially rewarding.

The transfer of subcultures from a supplied strain to third parties is strongly discouraged since the source is of fundamental importance in assessing authenticity. The passage of strains labelled with a service collection acronym and number from researcher to researcher is a dangerous practice, as contamination and mislabelling can occur lead-ing to considerable problems. When errors are discovered the original collection may, incorrectly, be regarded as responsible for the con-tamination. It is always advisable to obtain fresh subcultures of strains from a reliable service collection at the start of a research project, or to have their identity confirmed by an appropriate expert (see Chapter 5, Section 5.1).

In general, strains pathogenic to man, animals and plants will be supplied only to laboratories registered with the collection as equipped and competent to handle them. To avoid unnecessary delays in the dispatch of cultures, copies of appropriate certificates and licences should be sent to the collection when cultures are ordered.

3.1.2   Pricing

Pricing policy varies considerably between collections depend-ing on their funding arrangements and prescribed functions. Most are required to levy charges, and in the case of the larger service collections these form an important part of the financial base, without which services to the industrial and scientific communities could not be provided.

Almost all fungus culture collections receive substantial support from government departments, research councils, universities, or hospitals. As a result, they are able to supply cultures at prices substantially below full economic cost. In a major service collection using a wide range of modern preservation techniques, the full economic cost of a culture from the open collection is often 100–300% higher than the actual charges made to customers.

Cultures in 'open' collections may be differentially priced. Strains commonly requested for teaching purposes can be grown in bulk with consequent economies of scale; some collections (for example the American Type Culture Collection) have particularly long lists of such strains, while others do not regard it as a primary function to supply subsidised strains to schools and colleges. Again, sets of strains cited in official standards (for mould growth testing, for example, see Chapter 7, Section 7.2.2) are commonly ordered and can be produced in quantity and offered at reduced prices.

Variations in prices may also be related to the place of work of the purchaser. Higher prices are often charged for sales to industry than to educational establishments, research workers in universities or other publicly funded institutions. Reduced prices may be charged to organisations which directly contribute to the support of the collection.

In addition to advertised prices for cultures, some collections charge separately for the costs of postage, particularly where dispatch by air is involved. These costs can be high because of the need to pack the cultures carefully to avoid damage (see below). A service tax (such as the Value Added Tax, VAT, in EEC member countries) may also be chargeable on sales, depending on local taxation regulations. Usually charges such as VAT are only applicable to sales within the country in which the collection is based.

In contrast to the service collections, research collections are often able to make subcultures of their strains available at no cost whatsoever to fellow research workers. Generally, however, they do not have the facilities to supply a service to all customers on demand.

3.1.3   Packaging

Where possible, culture collections prefer to supply cultures to customers as freeze-dried samples in glass ampoules. Details of how to open the ampoules and revive the strains are usually provided. For strains that do not withstand the freeze-drying process (Chapter 4, Section 4.3.3), cultures grown on agar slants in glass Universal bottles are commonly supplied. Some collections use glass test-tubes with sterilised cotton wool plugs, although this may lead to increased risks of breakage and contamination.

When slants are dispatched, a solid agar medium optimal for the species is used. In the case of material sent to the tropics, media with increased amounts of agar (2.5–3%) is recommended to reduce the possibilities of the medium liquifying in transit.

Ampoules, bottles or tubes are individually wrapped and packed with cotton wool or, increasingly, polystyrene granules or shapes, within strong card, alloy, expanded polystyrene, or wooden containers (Fig. 3.1). Alloy screw-cap cylinders or solid balsa-wood bored with holes are used for packing glass test-tubes. The practice of placing agar cultures in small plastic bags or envelopes is unwise as they usually become contaminated. Regulations for the dispatch of infectious materials are described in Section 3.3 and safety regulations in Section 3.4.

Very little data on strains are usually sent with the cultures. In some cases only the accession number of the culture is provided. Some collections have facilities for producing adhesive labels in small type which allows inclusion of the name as well as the number of the fungus. For further information the catalogue entries for the strains must be consulted.

Fig. 3.1. Examples of packing materials suitable for the dispatch of ampoules, bottles, or tubes containing fungus cultures.

3.2 Deposit of cultures

3.2.1 *Acquisition policy*

The policy regarding the acquisition of cultures varies between culture collections. Acceptance of a culture for permanent preservation is a long-term commitment and also a long-term cost, not only because of preservation and storage costs, but because of the need for regular purity checks that form part of standard maintenance procedures. For these reasons collections may have to limit the numbers of strains they can accept. Further, collections may be restricted to certain groups of microorganisms, and research workers wishing to deposit strains in a service collection so that they will be available to others in the future should check that the acquisition policy of the collection includes categories covering their strains. Even where a strain is within the range of a collection, problems may arise if it can only be preserved by techniques not routinely available. Few collections are able to offer a fully comprehensive range of preservation methods and it is best to seek the advice of the collection before submitting cultures.

Culture collections are always interested in acquiring strains which extend the coverage of their fields of interest. Those serving systematics, for example, are usually pleased to receive strains of species not currently represented in their collections, or of which they have few recent isolates. Increasing numbers of collections welcome strains which are documented with regard to their biochemical, physiological, or genetic properties, although not all collections have facilities to check that these are retained during preservation and maintenance procedures. Fully documented strains are always of greater value (see below).

Some collections request subcultures of strains cited in scientific publications, and have staff who devote significant parts of their time to scanning literature for such isolates. Even if strains are already deposited in one service collection, research workers may welcome their strains being maintained in several places, so that they are more widely available and the risks of loss during preservation and maintenance are reduced.

The acquisition policy a collection adopts may be dictated by its funding body. The resultant variation is to be welcomed, as it means that the centres together constitute a much larger collective resource than would otherwise be possible. Although some idea of the role of a collection may be apparent from its name, as in the Fungal Genetics Stock Center, FGSC (Chapter 1, Section 1.5.3), this is not always so. The American Type Culture Collection, ATCC (Chapter 1, Section 1.5.2), for example, is not only concerned with cultures derived from nomenclatural types. The larger collections make their policy on acquisition clear in catalogues and other publications.

In the case of safe-deposit cultures (Chapter 7), the research worker determines which strains are deposited, subject to the collection having the appropriate facilities to maintain them.

It should be noted that *cultures containing mites present a hazard* to collections and will usually be destroyed on receipt.

3.2.2   Documentation

Cultures submitted to a culture collection for deposit should be as fully documented as possible. Data supplied with deposits should include as a minimum the scientific name of the strain, the host or substratum from which it was isolated, locality, name of isolator or depositor, and the reference number given to the strain by the sender. Other information that is useful to collections is summarised in Table 3.1. All of these categories of data are unlikely to be applicable or available for every strain, nor is the list of categories exhaustive.

With respect to the origin of the isolate, precise information is particularly welcome; for example, the part of host plants from which a culture came, whether the host was exhibiting any disease symptoms or was dead, is important information. In the case of soil isolates, the soil type, pH, horizon, and depth may be pertinent. In all cases, information on the method of isolation can be useful, particularly if the fungus has rarely been obtained in culture before.

Data on physiological and biochemical attributes are especially important to culture collections concerned with biotechnology, as there is an increasing interest in properties rather than in scientific names. Culture collections may wish to confirm these characters in the course of maintenance and preservation procedures, and consequently it is also helpful if the methods are cited by which particular activities or products were detected (e.g. thin-layer chromatography, high performance liquid chromatography) as the sensitivity of detection methods varies. This is of paramount importance in the case of secondary metabolites where the conditions under which compounds are formed can be crucial (e.g. liquid or solid cultures, incubation time).

Where strains have been cited in scientific papers, references or preferably reprints should be provided. Reprints are often filed with other papers in collections' records relating to a particular strain.

In order to assist depositors, the larger collections issue accession

Table 3.1. *Documentation of deposited cultures*

Name	Temperature range for growth
Identified by	Optimum growth temperature
Depositor's name	Light requirements
Depositor's address	Storage methods
Previous history	Metabolic products[a]
Other collections in which deposited	Enzymes produced[a]
Host substratum	Physiological properties[a]
Host symptoms	(e.g. growth rate)
Isolator's name	Antagonistic properties[a]
Isolate number	(to other organisms)
Growth media	Pathogenicity[a]
Isolation technique	Inhibitor's action
Country and Province	(antibiotics, poisons, etc.)[a]
Locality name	Special applications
National/UTM grid reference	References to citations in papers
Date of isolation	Restrictions on availability
Date subculture made	

[a] Plus methods used to determine these properties.

forms which depositors are invited to complete. Examples of these are given in Figs. 3.2–3.4. That drawn up by the World Data Center (see pp. 49–51) is illustrated in Fig. 3.5.

Fig. 3.2. Accessions form used by the American Type Culture Collection (ATCC).

**American
Type Culture
Collection**

12301 Parklawn Drive • Rockville, Maryland 20852 USA
Telephone: (301) 881-2600 • Telex ATCCROVE 908-768

Do not write in this box	
ATCC Number	62739
Accession Date	1/8/87
Date Received	12/19/86

COLLECTION OF FUNGI
To be completed by depositor of strain. Please print or type.

1. Scientific name and author ___Lecythophora mutabilis (van Beyma) W. Gams & McGinnis___

2. Source of species description ___Mycologia 75: (6), 977-987___

3. Classification: Order _____ Family _____

4. Name of other state ___Teliomorph not known___

5. Identified by ___L. Leightley, Queensland Dept. of Forestry, Australia___

6. Isolation and historical data:
   isolated by: ___L. Leightley___ date ___1979___ isolation number ___LQB 68-R___
   substrate or host ___Copper-chrome-arsenic treated Eucalyptus maculata Hook transmission pole at groundline region.___
   geographic source ___Tropical Australia___ significance of culture ___Tolerant to Copper, Chromium and arsenic and to creosote.___

   literature citation(s) for this strain ___Leightley, L.F (1980) cited in Hale, M.D. & Eaton, R.A. (1986) Trans. Br. Mycol. Soc. 86: 505-600.___

   location of herbarium specimens ___No known locations of herbarium specimens___

   cultures also deposited at ___CMI Ferry Lane, Kew, England, Numbers not yet allocated.___

7. Characteristics observed in culture **as deposited:**
   type of fruiting structures found ___Phialospores, dark chlamydospores.___
   factors affecting fruiting: temperature ___22 - 28°C___ light ___Unknown___
   preferred medium (attach formula) ___Oxoid CMA, 1.6% MA,AB cell___other___
   unusual maintenance requirements ___readily reisolated on 1.0% CuSO₄·5H₂O, 1.6% Malt Agar.___

8. Is this strain zoopathogenic? ___No___ If so, would you classify it as class 2, 3, or 4? ___
   (see reverse side for description of classes)

9. Is this strain phytopathogenic? ___No___ (Information required by Plant Quarantine Division, USDA) If so,
   a. Is the geographical distribution of this organism general, limited, or unknown (encircle)?
   b. Would you recommend that this strain be made available to any qualified investigator regardless of his location? ___
   c. If not, what limits would you place on the distribution of this strain?

10. Comments: ___Form and growth highly variable in culture dependent upon mode of culture. Adpressed on "minimal" media forming pigmented chlamydospores after a few weeks. More fluffy and pigmented, forming aerial hyphal aggregates on righer media.___

11. *I understand that this material is for deposit in the ATCC general collection. It will be examined, and if accepted by the ATCC, batches will be made and distributed to the scientific community for a fee to cover expenses.*
    ___5/2/86___   M.D. Hale
    Date   Signature of authorized individual
    Deposited on Behalf of ___Dr. M.D. Hale, Dept. Forestry & Wood Science,___

12. Depositor Address ___University College of North Wales, Bangor, Gwynedd LL57 2UW___

ATCC Form 1-F (1984)

Fig. 3.3. Accessions form used by the Centraalbureau voor Schimmelcultures (CBS).

CBS accession number (to be filled in by CBS)   CBS *445.86*

Data supplied on this form are of scientific importance; they will be recorded in the CBS files and be cited in part in the CBS List of Cultures. Reprints relating to this deposit will be highly appreciated. The strain will be regarded as free for distribution unless otherwise stated.

Scientific name and authority   *Chloridium paucisporum   Wang and Wilcox*

Deposited by (address)- *C.J.K. Wang-SUNY College of En-*   under number *BDD-22*
vironmental Science & Forestry, Syracuse, NY USA   *(Type culture)*
Identified by - *C.J.K. Wang*

Isolated by - *H.E. Wilcox*   date *March 1969*

Substrate and ecological data (please give latin name of host) *Isolated from short root*
of 3-yr.old seedling of Pinus resinosa *Ait.; forming ectendomycorrhizal association in*
Locality - *Syracuse, New York, USA*   P. resinosa, Picea rubens

Location of herbarium specimens *SYRF*

Preferred media - *Modified Melin-Norkrans agar, modified Hagem's agar, Potato dextrose*
Conditions for fruiting (temp., light) - *stationary Hagem's liquid medium*   *agar*
Special maintenance requirements

Morphological peculiarities - *sterile on solid media*

                                                           p.t.o.
Pathogenicity

Safety precautions required?

Decomposition of
Production of (enzymes, antibiotics, etc)

Resistance to

Strain also deposited at (possibly with accession nr) - *SYRF, ATCC, IMI, DAOM*

Literature citation(s) for this strain - *Wang, C. J. K. and H. E. Wilcox, 1985.*
Mycologia 77:951-958; Wilcox, H.E., Ruth Ganmore-Neumann, and C.J.K. Wang, 1974.
Can. J. Bot. 52:2279-2282

Additional data

                                           Signature

Fig. 3.4. Accessions form used by the CAB International Mycological Institute (IMI).

COMMONWEALTH MYCOLOGICAL INSTITUTE
Accession Form / History Sheet

FOR CMI USE ONLY

NAME & AUTHORITY	CMI ACCESSION NUMBER
TRICHODERMA VIRIDE Persoon	307987

PRINCIPAL SYNONYMS/NAME CHANGES

CMI ACCESSION DATE
11 08 86

IDENTIFIED BY:
M.A.J.Williams

TYPE OF DEPOSIT

DEPOSITOR: PLEASE FILL IN BELOW AS MUCH AS POSSIBLE

NAME OF ISOLATE:
TRICHODERMA VIRIDE   Persoon

DATE SENT TO CMI
24 07 86

NAME & ADDRESS OF DEPOSITOR:
Saddler, Forintek, Ottawa, Canada K1G 325, Canada

ISOLATE DESIGNATION

PREVIOUS HISTORY: (Other collections/owners/isolate designations)

OTHER COLLECTIONS WHERE HELD: (Give Collection Numbers)
Forintek 161P (D43)

ISOLATED FROM (Substratum/Genus & Species of Organism)
Indet. Host

ANATOMICAL PART/
SUBSTRATUM PART

GEOGRAPHICAL LOCATION: ISOLATED BY
Canada   K.A.Seifert

DATE OF ISOLATION

ISOLATION METHOD: (Soil Plate, Damp Chamber, Surface Sterilisation etc.)

SPECIAL FEATURES & USAGE (Metabolic products, Culture derived from Type etc.)
High extracellular cellulase activity

REFERENCES (Journal, Volume, page, year) Attach copies/reprints if possible.
Enzyme Microb. Technol. 4, 414-418. 1982

RECOMMENDATIONS FOR MAINTENANCE & PRESERVATION
Growth medium: Potato Carrot Agar
Incubation temperature: Room temperature
Incubation time: 7-14d
Light requirement/pH etc:
Subculative period: 4-6 weeks

FOR CMI USE ONLY

Please tick appropriate box(es)

Freeze drying	☐	Water storage	☐
Liquid nitrogen storage	☐	Soil storage	☐
Oil storage	☐	Silica gel storage	☐
Other – please specify			

Fig. 3.5. Accessions form SCC-4 used by the World Data Center for the storage of data on strains.

WFCC Form SCC-4

NOMENCLATURE DATA

Genus	Species (include Variety)	Type*
ASPERGILLUS	NIGER	F

= 100 Authority (Name and date)
van Tieghem

= 110 World Directory Collection Number: Collection Acronym IMI | Collection Accession Number 297707
214

HISTORY

= 120 Received as (Genus)
ASPERGILLUS

Species, variety
NIGER

= 125 Date received
07 Oct '86
Day Month Year

= 140 Strain Designation
As

= 130 Received from (Name and address of person or collection)
A.DELUCCA, USDA, Southern Regional
Research Centre, PO Box 19687, New
Orleans, LA 70179.

= 160 Strain designation
As

= 150 Who received it from (Name and address of person or collection)
–

= 170 Who received it from (Name and address of person or collection)
–

= 180 Strain designation
As

= 190 Also held in following permanent collections (include number)
–

ORIGIN

= 200 Isolated or derived from (If plant, animal or protist give genus and species name)
Stored cotton seed

= 210 Anatomical part (If applicable)
Seed

= 220 Isolated from (Country, nearest town and distance and direction there from — If possible give latitude and longitude in degrees, minutes, seconds and altitude)
USA New Orleans

= 230 Isolated by (Name)
A.Delucca

= 240 Date 86
Day Month Year

= 250 Identified by
Z.Lawrence.

= 260 Date
07 Oct '86
Day Month Year

MAINTENANCE

= 270 Method of Preservation (Please tick whichever is applicable) and = 280 Temperature of Storage

	°C	Agar Culture	°C
As Original Source Material			
Lyophilised Culture	°C	Liquid Nitrogen	°C

Other (Please specify)
under mineral oil at 15°C

= 290 Preservation suspending medium
1C% Skimmed milk + inositol

= 300 Growth Medium (Name and reference, give details on separate page)
MALT CZAPEK DOX AGAR

= 310 Temperature

= 320 Growth conditions (e.g. aerobic, anaerobic, special gas mixture, light)
AEROBIC DAYLIGHT

= 322 incubation time
7 days

= 327 Subculture period
60 days

SPECIAL FEATURES AND USAGE (e.g. for citric acid production, Type strain etc.)

= 330
Produces aurasperones A,B,C,D; iso aurasperone A;

flavasperone, rubrofusarin, fonsecin monomethylether

Enzyme activities: protease; amylase; cellulase;

Produces citric acid.

REFERENCES (Journal, volume, pages, year)

= 340
Applied and Environmental Microbiology 48 1-4 (1984)

= 990

* Type – B = bacteria, A = algae, F = fungus, Y = yeast, P = protozoa, AV = animal virus, BV = bacterial virus, PV = plant virus, IV = insect virus, L = lichen, TC = tissue cultures.

3.2.3   *Terms and conditions*

If a depositor wishes to place any restriction on the way a culture collection releases a deposited strain, he should make this clear when the strain is submitted. Where a safe-deposit is made, only the depositor has the right to the strain and the collection will make it available to him alone (Chapter 7); if it is a patent deposit special conditions apply (Chapter 6). However, in all other cases where the strain is included in the 'open' collection, no restriction is placed on potential purchasers, although differential pricing policies may operate (see Chapter 3.1.2). Information on strains in the 'open' collections are included in printed catalogues and may also be included in computerised databases (Chapter 2).

Culture collections treat sales in confidence and do not normally inform depositors who has purchased subcultures of their strains. The only exception to this is in the case of requests for patent deposit strains (Chapter 6).

In return for depositing strains in the 'open' collections, some collections supply subcultures to the original depositor free of charge, up to once or twice a year. Further, certain culture collections anxious to encourage acquisitions permit depositors to receive an equal number of other strains from their 'open' collection in free exchange if requested at the time of deposit.

3.3   **Postal regulations**

The distribution of cultures, both domestically and overseas, is essential to further scientific endeavour world-wide. However, the shipment of microbial cultures subjects culture collections, as well as individual scientists, to a complex set of laws that have been designed to protect humans, plants and animals from disaster through accidental release or introduction of infectious agents. Most countries have laws concerning both the import and the export of microbial cultures, and frequently the states and provinces of these countries also regulate movement of pathogenic materials. Accordingly, those who ship or import cultures need to ensure compliance with these laws and regulations.

In addition to proper packaging to avoid risk of injury to postal staff or damage to other mail (see Section 3.1.3.), living and preserved specimens are usually only despatched in response to orders received on official stationery. Organisms used in the biocontrol of insect pests should only be sent to institutions registered with the collection as being able to handle them safely. Where different rates of postage exist, specimens are usually sent by first class mail, or the equivalent service, and marked 'URGENT – BIOLOGICAL SPECIMENS'. Special distinctive labels are issued in many countries.

With regard to transport by air, the International Civil Aviation Organisation (ICAO) and International Air Transport Association (IATA) have reached agreement with curators of culture collections on procedures to be followed. Information is published by ICAO (1985) (see Chapter 3.4.3) providing details of the designation of dangerous goods, with packing instructions, which may be sent by air. A list of operators holding general permission to carry such goods by air is also available. This information is continually updated and reviewed.

Infectious perishable biological substances (IPBS), that is organisms or toxins known or suspected to cause disease in man or animals, must be packed according to national and international postage regulations and sent by registered post with an appropriate customs declaration label. Authorisation for supply must be obtained in advance from the national postal headquarters. IPBS material must be packed inside two watertight packages with absorbent internal packing between. Packages containing infectious or non-infectious material and not complying with the appropriate regulations may be intercepted and destroyed by the postal authorities. Class IV pathogens in the USA require a permit.

In the UK conditions for the acceptance of Infectious Perishable Biological Substances (IPBS) must be sought from the Post Office Authorities prior to despatch, usually from a nominated Crown Post Office. Despatch of perishable biological substances in the overseas post is restricted to those countries whose postal administrations are prepared to admit such items. A list of countries accepting these items appears in the *Post Office Guide*, available from certain HMSO Offices (see Chapter 3.4.3). Senders are strongly recommended to ascertain from the addressee before despatch that the goods will be acceptable in the country of destination.

Customs declaration forms are required for all overseas despatches and import licences will be required for plant pathogenic fungi (see Section 3.5 below).

In the UK the information on legislation governing the import and export of infectious biological substances can be obtained from either the

Health and Safety Executive (HSE), Magdalen House, Stanley Precinct, Bootle, Merseyside, or the Ministry of Agriculture, Fisheries and Food (MAFF), Hook Rise South, Tolworth, Surbiton, Surrey.

In the USA the importation of infectious biological substances capable of causing human disease is controlled by the Public Health Service Foreign Quarantine Regulations (42 CFR, Section 71.156). Importation permits, conditions of shipment and handling procedures are issued by the Centers for Disease Control. Similar restrictions and conditions for the importation of animal pathogens other than those affecting humans are subject to US Department of Agriculture Regulations, Hyattsville, Maryland. The ATCC (1986) (see Further reading) provides a summary of regulations affecting US scientists. The National Institutes of Health (NIH) have published a Laboratory Safety Monograph which provides additional information.

It is impossible to detail procedures for all countries here, but reference to the reading list (Section 3.4.3) should enable the reader to obtain the necessary information. Information on appropriate packing material and suppliers can be obtained through the service culture collections listed in Chapter 1. A number of culture collections provide in their catalogues useful information on aspects of safety, or have advisory leaflets available for distribution. There is no doubt that continued interest and concern will be expressed by both the scientific community and the general public about the safety aspects surrounding microbiological agents. Safety professionals and scientists alike have an obligation to ensure that all microbiological agents are handled in a safe and responsible manner.

3.4   Health and safety

3.4.1   *Health considerations*

While most filamentous fungi are not known to have any adverse effects on human health, some can attack man directly, initiate allergic responses, or produce highly toxic metabolites termed mycotoxins. Microfungi should always, therefore, be handled with care.

Safety aspects are discussed briefly here, but further information on fungi and related medical problems may be obtained from Emmons *et al.* (1977), Howard (1983–5), Evans & Gentles (1985) and McGinnis (1980). The literature on medical mycology is abstracted in the quarterly *Review of Medical and Veterinary Mycology* (1943 onwards; also on-line from 1973 in CAB ABSTRACTS).

Human pathogens. Systematic mycoses, invasions of living tissues, are caused by several fungi, although none are listed above Level 3 containment (Section 3.4.2). Most serious pathogens are *Ajellomyces capsulatus* (*Histoplasma capsulatum*, histoplasmosis), *Coccidioides immitis* (coccidioidomycoses) and *A. dermatidis* (*Blastomyces*, blastomycosis); but opportunistic species, particularly members of the Mucorales and certain *Aspergillus* species (e.g. *A. fumigatus*), can cause problems in immunologically compromised subjects.

Dermatophytes, attacking skin and nails, are much more widespread, and include the ring-worm fungi; the *Microsporum* and *Trichophyton* anamorphs of *Nannizia* and *Arthroderma* respectively, cause diseases referred to as 'tinea'. Saprobic fungi (e.g. *Exophiala jeanselmei*, *Neotestudina rosatii*, *Pseudallescheria boydii*), including *Sporothrix schenckii* that causes sporotrichosis, can enter through wounds and cause subcutaneous problems, including mycetomas. A wide range of fungi is involved in such cases and the damage caused varies considerably. In severe cases, fungi can spread through the vascular system to other parts of the body.

Fungi causing allergic reactions. Fungal spores give rise to allergic (hypersensitivity) reactions in some individuals. A wide range of species, all producing vast numbers of airborne spores, have been implicated in such responses, and the genera involved include species of *Aspergillus*, *Botrytis*, *Cladosporium*, *Helminthosporium*, *Mucor*, *Penicillium*, *Puccinia*, *Rhizopus*, *Serpula* and *Ustilago*. The allergic responses initiated may include asthmatic breathing conditions and can be severe in susceptible individuals. Laboratory practices consequently need to minimise the risk of large numbers of spores being released into the atmosphere. Staff particularly prone to such conditions are well advised to avoid lengthy periods of work with abundantly sporing filamentous fungi.

Mycotoxins. About 300 fungal metabolites are known to be toxic and/or carcinogenic to man; many fungal products have not, however, been tested for toxicity or carcinogenicity, and new mycotoxins are continually being reported. Amongst the most dangerous are aflatoxins (*Aspergillus flavus*), the trichothecenes (*Fusarium* species) and ochratoxin-A (e.g. *Aspergillus ochraceus*, *Penicillium viridicatum*). Infection most commonly results from the ingestion of infected foodstuffs. Symptoms may include severe alimentary disturbances, eczema, abortions, or haemorrhages, and death may result.

Common mycotoxins are listed in Table 3.2. Particular care should be taken when handling fungi reported to produce these compounds.

Poisoning can also be caused by eating incorrectly identified or incorrectly prepared larger fungi (Hymenomycetes), such as certain species of *Amanita, Gyromitra, Inocybe, Panaeolus* and *Psilocybe*.

For further information on toxic fungi and the hazards they constitute, Cole & Cox (1981) and Moreau (1979) should be consulted.

3.4.2   *Safety regulations*

Since the introduction of the Occupational Safety & Health Act, 1970, in the USA and the Health & Safety at Work Act in 1974 in the UK, a spate of guidelines, regulations or recommendations have been produced pertaining to the handling of microorganisms in research and production establishments, hospitals and educational departments. There has been a conscious attempt to ensure that acceptable safety procedures are established and maintained at all places of work. In the USA in 1984, the Centers for Disease Control and National Institutes of Health jointly prepared *Biosafety in Microbiological and Biomedical Laboratories*, which describes combinations of standard and special microbiological practices, safety equipment and facilities that constitute Biosafety Levels 1–4 recommended for working with a variety of infectious agents. The four classes and examples of organisms in each class are listed in Table 3.3. Table 3.4 outlines recommended laboratory handling procedures for infectious agents.

Table 3.2. *Examples of toxic secondary metabolites (mycotoxins) produced by filamentous fungi*

aflatoxins	ochratoxin-A
amanitins	patulin
citreoviridin	penitrem-A
citrinin	psilocybin
cochloidinol	roridins
coprine	rubratoxins
cyclopezionic acid	saframine
ergotamine	sporidesmin
gyromitrin	sterigmatocystin
leutoskyrin	tenuazonic acid
lysergic acid	trichothecenes
muscarine	zearalenone

In the UK the Health & Safety Executive offers an advisory service to the community in all matters of safety and has inspectoral powers to enforce the law in instances where the Health & Safety at Work Act is breached. The latest guidance on the categorisation of pathogens is to be found in a report produced by the Advisory Committee on Dangerous Pathogens (ACDP) 1984, *Categorisation of Pathogens according to Hazard and Categories of Containment*. Bacteria, chlamydiae, rickettsiae, mycoplasmas, fungi, viruses and parasites are clearly categorised according to the hazard they present to workers and the community, and four hazard groups are identified (1–4). Information is given on the degree of containment and protective clothing which should be applied during the handling of such organisms in the laboratory, including requirements for animal containment.

Table 3.3. *Classification of microorganisms according to biological hazard and their shipping requirements*[a]

Class I	Agents of no recognised hazard under ordinary conditions
Examples	*Saccharomyces cerevisiae, Trichoderma reesei, Lactobacillus casei*
Shipping	Culture-tube in fiberboard or other container. Permits as required
Class II	Agents of ordinary potential hazard
Examples	*Aspergillus fumigatus, Candida albicans, Cryptococcus neoformans, Staphylococcus aureus*
Shipping	Culture-tube wrapped in absorbent material, placed in metal screw-cap can, placed in fiberboard container. Permits as required
Class III	*Pathogens involving special hazard*
Examples	*Coccidioides immitis, Ajellomyces capsulatum, Bacillus anthracis, Yersinia pestis*
Shipment	Culture-tube heat sealed in plastic, wrapped in absorbent material, placed in hermetically sealed can, placed in sturdy cardboard box. Permits as required. Etiologic agent warning label necessary
Class IV	*Pathogens of extreme hazard*
Examples	*Arthroderma simii, Pasteurella multocoida,* certain animal/plant viruses
Shipment	Culture-tube heat sealed in plastic, wrapped in absorbent material, placed in hermetically sealed can, placed in sturdy cardboard box. Required permits. Etiologic agent warning label necessary

[a] US Department of Health, Education and Welfare, 1972; US Department of Health and Human Services, Public Health Service, 1983.

Fungal examples are:

  Blastomyces dermatidis   category 3
  Histoplasma species   category 3
  Aspergillus flavus   category 2
  Trichophyton species   category 2
  Sporothrix schenckii   category 2

For fungi in containment category 2, the requirements are for dedicated wash-hand basins in each laboratory and autoclave facilities in each

Table 3.4. *Summary of recommended biosafety levels for infecting agents[a]*

Biosafety level	Practices and techniques	Safety equipment	Facilities
1	Standard microbiological practices	None: primary containment provided by adherence to standard laboratory practices during open-bench operations	Basic
2	Level 1 practices plus: laboratory coats; decontamination of all infectious wastes; limited access; protective gloves and biohazard warning signs as indicated	Partial containment equipment (i.e. Class I or Class II Biological Safety Cabinets) used to conduct mechanical and manipulative procedures that have high aerosol potential that may increase the risk of exposure to personnel	Basic
3	Level 2 practices plus: special laboratory clothing; controlled access	Partial containment equipment used for all manipulations of infectious material	Containment
4	Level 3 practices plus: entrance through change room where street clothing donned; shower on exit; all wastes decontaminated on exit from facility	Maximum containment equipment (i.e. Class III biological safety cabinet or partial containment equipment in combination with full-body, air-supplied, positive-pressure personnel suit) used for all procedures and activities	Maximum containment

[a] US Public Health Service, 1983.

suite; safety cabinets are generally optional. In containment category 3 the laboratory site must be partially isolated, able to be sealed for fumigation, and have inward airflow or negative pressure ventilation (optionally through use of a safety cabinet). No fungi are in category 4, which includes the Lassa fever virus.

The American and British guidelines vary slightly but both have been accepted by the World Health Organisation. Other countries have developed similar systems or use those operating in the USA or the UK.

The control of genetic manipulation experiments in the UK is the responsibility of the Advisory Committee on Genetic Manipulation, using its own classification of experiments 1–4. Guidance is offered in a series of newsletters which are constantly updated.

The American control system is somewhat more complicated but all federal agencies that fund research related to biotechnology adhere to the policy that research in this field must conform to National Institutes of Health Recombinant DNA Guidelines. Other countries engaged in recombinant DNA techniques have local guidelines or use the American or British guidelines.

3.4.3   *Further reading (in chronological order) on health and safety issues*

The Occupational Safety and Health Act of 1970. United States Occupational Safety and Health Administration. Public Law 91-596. 91st Congress S.2193. USA.

US Department of Health, Education and Welfare (1972). *Classification of Etiologic Agents on the Basis on Hazard.* Center for Disease Control, Atlanta, Georgia 30333, USA.

The Health and Safety at Work, etc. Act 1974, HMSO, London.

List of Dangerous Goods and Conditions of Acceptance by Freight Train and by Passenger train, BR. 22426 (1977 under revision), available from Claims Manager, British Rail Board, Marylebone Passenger Station, London NW1 6JR, UK.

Laboratory Safety Monograph: A Supplement to NIH Guidelines for Recombinant DNA Research (1978). National Institutes of Health, Bethesda, Maryland, USA.

Vorläufige Empfehlungen für den Umgang mit pathogenen Mikroorganismen und für die Klassifikation von Mikroorganismen und Krankheitserregern nach dem im Umgang mit ihnen auftretenden Gefahren. *Bundesgesundheitsblatt* **23**, 347–59 (1981).

US Department of Health and Human Services, Public Health Service (1983). *Biosafety in Microbiological and Biomedical Laboratories.* National Institutes of Health, Bethesda, Maryland, USA.

Categorisation of Pathogens according to Hazard and Categories of Containment (1984). Advisory Committee on Dangerous Pathogens, HMSO.

Biosafety in Microbiological and Biomedical Laboratories (1984). US Department of Health and Human Services, Public Health Services, Centers for Disease Control and National Institutes of Health, HHS publication No. (CDC) 84-8395.

The Air Navigation (Dangerous Goods) Regulations (1985). English language edition of the International Civil Aviation Organisation. Technical Instructions for the Safe Transport of Dangerous Goods by Air (DOC 9284-AN/905).

American Type Culture Collection (1986). *Packaging and Shipping of Biological Materials at ATCC*, Rockville, Maryland, USA.

LAV/HTLV III – the causative agent of AIDS and related conditions. Revised guidelines (1986). Advisory Committee on Dangerous Pathogens, DHSS. Health Publications Unit No. 2 Site, Manchester Road, Heywood, Lancs. OL10 2PZ, UK.

Undated references

Advisory Committee on Genetic Manipulation. Guidelines and newsletters. ACGM Secretariat, Baynards House, 1 Chepstow Place, London W2 4TF, UK.

Conditions of Supply of NCTC Cultures: Hazardous Pathogens. Public Health Laboratory Service, Central Public Health Laboratory, 61 Colindale Avenue, London NW9 5HT, UK.

Importation Permits available from Centers for Disease Control, 1600 Clifton Road NE, Atlanta, Georgia, USA.

Public Health Service Foreign Quarantine Regulations, 42 CFR Section 71-156.

The Post Office Guide, available from certain HMSO Offices, including 49 High Holborn, London WC1V 6HB, UK.

3.5   Quarantine regulations for plant pathogenic fungi

Fungi are the major cause of plant diseases in the world. Some are widely distributed throughout crop-growing regions, but many are restricted in their occurrence. Legislation has been introduced to endeavour to eliminate the risk of a plant pathogenic fungus inadvertently being introduced into a country from which it has not hitherto been recorded. Quarantine regulations relate to plants and seeds, as well as living cultures of the disease-causing fungi, but only the last normally concern the biotechnologist interested in fungi.

It is important for biotechnologists to remember that many of the genera of fungi they work with include plant pathogenic species (e.g. *Fusarium, Verticillium*).

The major part of the legislation relating to quarantine for plant disease is concerned with phytosanitary regulations which apply to plant material. A valuable summary of these is provided by Johnston & Booth (1983), and more extensive information is given in Hewitt & Chiarappa (1977) and various publications of the Food and Agriculture Organisation (FAO). The *FAO Plant Protection Bulletin* is a valuable source of recent information, and world distributions of some 600 plant diseases are given in the twice-yearly *CMI Distribution Maps of Plant Diseases*.

National procedures vary considerably with regard to regulations for the import of plant pathogenic fungi. Most maintain a list of species which cannot be imported without a licence. Licences are generally issued by a special section of the ministry of agriculture on a species-by-species basis. The issue of a licence usually involves inspection of the receiving research institute's laboratories to ensure that they are adequately equipped to contain the fungus within buildings or glass-houses. The inspectors will normally need to be satisfied as to the procedures for opening packages, safe laboratory practices and the effectiveness of methods for the disposal of unwanted material (e.g. autoclaving).

Major service culture collections generally work closely with quarantine departments which inspect their facilities from time to time. If they are satisfied that the laboratory procedures will not lead to the release of the fungus into the surrounding area they may then issue a licence to import *any* plant pathogenic fungi. Although licensed collections can then receive and maintain cultures of plant pathogenic fungi from anywhere in the world, they are not authorised to distribute subcultures to any third party which has not itself been authorised by the ministry to receive them.

Many collections indicate in their catalogues strains which they cannot distribute to unlicensed institutions or individuals. If such a strain is required, the intending purchaser should obtain the appropriate licence or authorisation and enclose a copy of it when placing an order. If this procedure is not followed, any collection which services the order runs the risk of losing its own licence to handle such strains, and this could seriously affect its ability to act as a comprehensive resource centre.

3.6   Security

Culture collections take steps to ensure that their cultures and the records relating to them are secure. Even where fire precautions and building security are to high specifications, as is appropriate to international germ plasm centres, the total loss of a collection due to unforeseen circumstances remains a possibility. The major service collections consequently maintain a duplicate set of subcultures or ampoules in another building, preferably below ground level and at a spatially distinct site. When such a set is kept, then in the event of a major disaster, the collection could be reconstituted.

Records are a vital complement to the collection and so also merit careful preservation. Paper documents can be microfilmed at regular intervals, as is the practice with the entire herbarium (dried specimens) record books at the CAB International Mycological Institute. Now that collection records are increasingly stored in computers, this aspect of security is much easier, as back-up copies of disks can be made routinely. Both microfilms and back-up disks can be housed in a different building or site.

Depositors frequently send subcultures to more than one service collection to minimise risks of loss due to death or contamination of the strains. Since difficulties can arise through transport or accessioning procedures, it is prudent for depositors to retain a live subculture until the collection to which it has been sent confirms that it has received and incorporated the strain.

Where strains exhibit unusual properties that a collection may not have the capacity to check, the depositor should consider asking the collection to return a fresh subculture or ampoule *after* preservation in order to confirm that the critical properties have not been lost during the transfers necessary in the preservation process.

4

Culture and preservation

D. SMITH

4.1 Introduction

It is essential that the most appropriate growth conditions and preservation techniques are used to ensure the viability, purity and stability of maintained microorganisms. Ideally, the methods used should retain all characteristics throughout storage.

Many filamentous fungi grow on culture media and can be kept viable by periodic transfer. However, the properties of fungi in culture may be unstable through loss of plasmids, spontaneous mutations or genetic recombination (due to the presence of heterokaryons, the parasexual cycle or normal sexual events). These phenomena can result in modification of a strain's characteristics, and conditions of preservation and storage should be selected to minimise the risk of such changes.

4.2 Growth of cultures

The major factors affecting growth are medium, temperature, light, aeration, pH and water activity.

4.2.1 Media

The growth requirements for fungi may vary from strain to strain, although cultures of the same species and genera usually grow best on similar media. The source of isolates can give an indication of suitable growth conditions, thus isolates from jam can be expected to grow well on high-sugar media, species from leaves may sporulate best in light, those from marine situations may require salt and those from deserts and the tropics, high growth temperatures.

Cultures are usually best grown on agar slopes in test-tubes or culture bottles. A list of recommended media and growth temperatures for common species is given in Smith & Onions (1983). Details of the most frequently used media are given in the Appendix at the end of this book. The majority of fungi can be maintained on a relatively small range of media. However, some fungi deteriorate when kept on the same medium for prolonged periods, so different media should be alternated from time to time.

4.2.2   Temperature

The majority of filamentous fungi are mesophilic, growing at temperatures within the range of 10–35 °C, with optimum temperatures between 15 and 30 °C. Some species (e.g. *Aspergillus fumigatus, Talaromyces avellaneus*) are thermotolerant and will grow at higher temperatures, although they are still capable of growth at room temperature. A small number (e.g. *Chaetomium thermophilum, Penicillium dupontii, Thermoascus aurantiacus*) are thermophilic and will grow and sporulate at 45 °C or higher, but fail to grow below 20 °C. A few fungi (e.g. *Hypocrea psychrophila*) are psychrophilic and are unable to grow above 20 °C, while many others (e.g. a wide range of *Fusarium* and *Penicillium* species) are psychrotolerant and are able to grow both at freezing point and at room temperature.

4.2.3   Light

Many species grow well in the dark, but others prefer daylight and some sporulate better under near ultraviolet light (see Section 4.2.7). Most leaf- and stem-inhabiting fungi are light sensitive and require light stimulation for sporulation.

4.2.4   Aeration

Nearly all fungi are aerobic and, when grown in tubes or bottles, obtain sufficient oxygen through cotton wool plugs or loose bottle caps. Care should be taken to see that bottle caps are not screwed down tightly during the growth of cultures. A few aquatic Hyphomycetes require additional aeration, by bubbling air through liquid culture media for example, to enable normal growth and sporulation to occur.

4.2.5   pH

Filamentous fungi are variable in their pH requirements. Most common fungi grow well over the range pH 3 to 7, although some can grow at pH 2 and below (e.g. *Moniliella acetoabutans, Aspergillus niger, Penicillium funiculosum*).

4.2.6   Water activity

All organisms need water for growth, but the amount required varies widely. Although the majority of filamentous fungi require high levels of available water, a few are able to grow at low water activity (e.g. *Eurotium* species, *Xeromyces bisporus*). Some of those which occur on preserves or salt fish will only grow well on media containing high concentrations of sugar or salt. These fungi are referred to as xerophiles or halophiles.

4.2.7   Near ultraviolet light (black light)

Fungi which require near ultraviolet light (wavelength 300–380 nm) for sporulation must be grown in plastic Petri dishes or plastic universal bottles for 3–4 days before irradiation. Glass is not suitable, as it may be opaque to ultraviolet light. Rich growth media should be avoided, as they may give rise to excessive growth of mycelium; nutritionally weak media such as potato carrot agar (PCA) are more suitable for inducing sporulation.

In one irradiation system three 1.22-m fluorescent tubes [a near ultraviolet light tube (Phillips TL 40 W/08) between two cool white tubes (Phillips MCFE 40 W/33)] are placed 130 mm apart. A time switch gives a 12-hour on/off cycle. The cultures are supported on a shelf 320 mm below the light source and are illuminated until sporulation is induced (Smith & Onions, 1983).

4.2.8   Control of mites

Mites, particularly species of *Tyrophagus* and *Tarsonemus*, sometimes contaminate fungus cultures. They form a very serious problem in fungus culture collections, since they transfer spores from culture to culture, causing severe cross-contamination that may be irremediable.

To prevent mite invasion, all work surfaces must be kept clean and cultures protected from aerial infestation by storage in cabinets or incubators. To clean work surfaces, wash with 70% ethanol or the non-fungicidal acaricides 0.5% (w/w) Tedion V-18 (Mi-dox Ltd) or 0.2% v/v Actelic (ICI plc). Mites can be detected by scrutiny of cultures at twice weekly intervals. The mites can be seen as white objects, just detectable with the naked eye. Ragged colony margins or growth of contaminant fungi or bacteria forming trails may denote their presence. If mites are detected, contaminated cultures should be destroyed by steam, alcohol or autoclaving.

If mite-infested cultures cannot be reisolated, they can be stored at $-18\,°C$ for 1–3 days before being subcultured. This procedure will kill both eggs and adult mites. Fungi which would not survive short-term cold storage may be covered with a layer of mineral oil and subcultured after 24 hours, although this procedure does not kill the eggs.

A cigarette paper fastened on the necks of universal bottles or tubes provides an effective barrier against mites. The cigarette papers are sterilised in propylene oxide overnight and attached by rotating the bottle neck in copper sulphate/gelatin glue (20 g gelatin, 2 g copper sulphate in 100 ml distilled water) and placing the paper on the neck. The excess paper can then be flamed away.

Storage of cultures by freeze-drying and on silica gel excludes the possibility of mite infestation and temperatures below $8\,°C$ immobilise them.

4.3   Scale-up

There are many industrial processes in which the biochemical properties of fungi are harnessed for the benefit of man (see Chapter 1, Section 1.2). To scale up from growth of fungi on agar media in Petri dishes or bottles in the laboratory to the volumes necessary on an industrial scale presents many problems. The sterilisation of media on a large scale is expensive. An abundant supply of sterile air must be provided to fermenters, or the fermentation must be carried out in shallow layers of liquid with free aeration. In the latter case protection from contamination is necessary and requires expensive equipment and expert supervision. Again, the fungus strain may not perform in the large-scale production as it does in laboratory or even in pilot tests. The problems involved in 'scale-up' have been reviewed (Charles, 1985). However, citric acid is produced commercially using strains of *Aspergillus niger* and industry has overcome many of these problems.

Most of the major advances in antibiotic yield have been as a result of the genetic modification of the production strains. These manipulated strains must then be preserved to prevent culture degeneration. Once this has been achieved, the fungus must be grown in bulk under suitable conditions and inoculated into the production medium, avoiding stresses that may prevent the fungus functioning.

The preparation of a large inoculum requires growing the organism in several stages of ever-increasing volume to obtain an inoculum large enough for the final commercial stage. The inoculum medium should be designed for rapid cell growth and not for product formation. After growth, cells must be concentrated so that the inoculum does not dilute the production medium excessively, change the pH or add unwanted metabolic products when added to the substrate. The growth conditions used must be selective for the production strain and not give mutants in the population a growth advantage. Mutants used as production strains may be unstable and easily revert to parental strains, and stock cultures must be preserved in such a way that the least number of back mutants form part of the inoculum.

Spores from many fungi float and require a wetting agent such as sodium lauryl sulphonate to get them into suspension. To produce a mycelial mat on the surface of the growth medium, essential for some fermentations, it may be necessary to spray the spores or float them onto the surface.

Contamination is a major problem during bulking of inocula, and during the process this can cause serious problems. It is essential that throughout all transfers and growth the fungus inoculum and growth media are contaminant free. Further information on industrial 'scale-up' can be obtained from Casida (1964).

4.4   Preservation techniques
The preservation of filamentous fungi has been well covered in Smith & Onions (1983) and Kirsop & Snell (1984). Reference should be made to these publications for further experimental details of the methods discussed below.

4.4.1   *Subculturing*
Many fungi can be maintained for years by growth on suitable media, providing they are transferred to fresh media from time to time. Successful maintenance is dependent upon transfer from well-developed parts of the culture, taking care to ensure that contaminants or genetic variants do not replace the original strain.

On agar. Cultures on agar slants may be kept at room temperature. Shelf-life is usually between 2 and 6 months, although some fungi, such as certain water moulds, require transfer every 1–2 weeks (Table 4.1) while others remain viable for several years. Storage in refrigerators at 4–7 °C enables the period between transfers to be extended and, as mites do not invade cultures at these temperatures, this practice is recommended (Fig. 4.1).

On agar under oil. The period between transfers can also be increased by covering a culture growing on an agar slant (30° to the horizontal) with 10 mm of medicinal paraffin (Buell & Weston, 1947). This reduces metabolism by restricting the oxygen supply and also prevents dehydration. If the oil is deeper than 10 mm, the fungus may not receive sufficient oxygen and may die, while if the depth is less, exposed mycelium or agar on the sides of the container may allow moisture to evaporate and the culture to dry out.

Storage under oil has been used successfully with many groups of fungi and significantly extends the storage period between transfers to fresh media (Table 4.2; Fig. 4.2). The low cost and simplicity of this technique make it very useful for laboratories with limited resources.

In water. Some aquatic and terrestrial fungi can be successfully stored in water (Table 4.3). Spores or hyphae removed from the surface of agar media or blocks cut from the growing edge of colonies can be placed in sterile water, sealed in tubes or bottles and stored at temperatures between 4 °C and room temperature.

Growth may sometimes occur during storage in water, but the fungus eventually becomes dormant. Successful storage of Entomophthorales,

Table 4.1. *Shelf-life of fungi stored on agar at room temperature or under refrigeration*

Fungus	Storage temperature (°C) and conditions	Period before transfer[a]
Ascomycotina	4–7	1 year
	room	2–6 months
Basidiomycotina	4–7	1 year
	room	2–6 months
Chytridiomycetes	16	2–3 months
Deuteromycotina	5	1 year
Dermatophytes	18–25 on hair	2–3 months
Oomycetes	4	1 year
	16	2–3 months
Rhizopus and other low-temperature-sensitive fungi	16, 60% relative humidity	6 months
Zygomycotina	4–7	1 year
	room	1–2 months

[a] This period is not necessarily the maximum storage period.

Fig. 4.1. Cultures stored on agar slants at 5 °C at the Centraalbureau voor Schimmelcultures (CBS).

Ascomycotina, Hymenomycetes, Gasteromycetes and Hyphomycetes has been claimed (Ellis, 1979), but tolerance of this treatment is very variable. After 2 years, viability or infectivity may be lost.

Advantages and disadvantages. For small collections or for relatively short-term preservation, subculturing is an effective method of culture maintenance. However, growing cultures are particularly subject to airborne

Table 4.2. *Shelf-life of fungi stored on agar under oil at room temperature*

Fungus	Period before transfer (years)
Alternaria[a]	0.5
Aspergillus spp.	20–32
Basidiomycetes	10
Chlamydomyces palmarum	32
Corticium spp.	20–32
Drechslera portulacae	32
Fusarium spp.[a]	0.5
Mortierella alpina	12
Nectria spp.	32
Mucor spp.	20
Penicillium spp.	20–32
Pythium spp.	3
Trichoderma harzianum	27
Trichophyton spp.	12
Verticillium spp.	32
Volutella ciliata	32

[a] Retained pathogenicity.

Fig. 4.2. Cultures stored under mineral oil and lyophilized at the CAB International Mycological Institute (IMI).

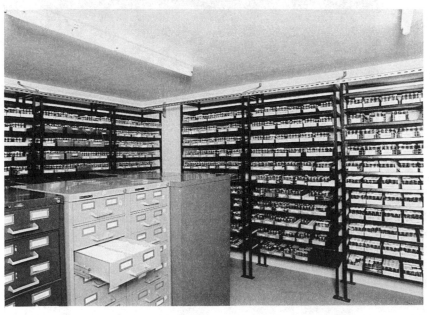

and mite contamination. If the contamination is not too great, it may be possible to reisolate the fungus from the infected culture, but this may prove difficult.

The preservation techniques so far discussed (Section 4.4.1) allow the organism to grow and metabolise and are used to extend the period between transfers to new media. Storage of fungi in the refrigerator at 4–7 °C slows down the rate of metabolism and increases the period between transfers to fresh media. Some fungi are sensitive to storage at these temperatures, particularly thermophiles and many Mastigomycotina. However, fungi that are not sensitive will have almost double the shelf-life using this method. Overpacking of refrigerators can cause a build-up of condensation which may promote cross-contamination.

One of the most important disadvantages of the oil storage technique is that some fungi may adapt to it, or spontaneous mutants may grow better in the conditions provided (Onions, 1983). Some cultures grow better after recovery from oil than after maintenance by more frequent transfer techniques (Table 4.4).

Table 4.3. *Shelf-life of fungi stored in water*

Fungus	Inoculum	Storage temperature (°C)	Period before transfer (years)
Cladosporium mansonii	spores and hyphae	room	1
Colletotrichum	agar	1	0.5
Conidiobolus spp.	agar blocks	25	1.75
Ectomycorrhizal fungi	agar blocks	5	1–3
Entomophthora coronata	agar blocks	25	1.75
Epidermophyton floccosum	spores and hyphae	room	1
Geotrichum spp.	spores and hyphae	room	1
Hymenomycetes	agar blocks	25	0.4–1.75
Monilia spp.	spores and hyphae	room	1
Mucorales	agar blocks	25	0.4–1.75
Nectria radicicola	agar blocks	25	1.75
Phytophthora spp.	agar blocks	15	2–3
Pythium spp.	agar blocks	15	2–3
Saprolegnia furcata	mycelium on tellurite agar blocks	4	8
Trichophyton rubrum	spores and hyphae	room	1

Table 4.4. *Strains retaining characteristics after recovery from oil storage after 20 years*[a]

Name	Retained characteristic
Aspergillus avenaceus	production of abundant sclerotia
Botryosphaeria obtusa	production of ascospores
Penicillium adametzii	profuse sporulation
P. asperum	production of sclerotia
P. lapidosum	sclerotia and sporulation
P. oxalicum	profuse sporulation
P. waksmanii	profuse sporulation

[a] These characteristics were lost during more frequent serial transfer.

In one study (Onions, 1977), 79% of 58 strains belonging to 23 genera survived for 23 years when stored under mineral oil. Although most isolates were recovered without morphological change, some needed further transfers before a healthy culture was obtained. Sectoring of the colony into different morphological types was observed in some strains.

Storage temperatures below 8 °C are strongly recommended both to extend transfer time and avoid predation by mites. In general, as serial transfer is susceptible to contamination and can encourage variation (Onions, 1983), other preservation methods should be sought for long-term storage.

4.4.2   Drying

Most fungal spores have a lower water content than vegetative hyphae and are able to withstand desiccation. Dehydration suspends metabolism and only when water is restored will the fungus revive and grow. Several preservation techniques rely on drying. For example many *Aspergillus* and *Penicillium* species survive air-drying (Smith & Onions, 1983). *Allomyces arbuscula* has been reported to survive for up to 17 years (Goldie-Smith, 1956). The spores of vesicular-arbuscular endo-phytes (e.g. *Glomerus, Acaulospora, Gigaspora*) have been preserved by drying at 22 °C under vacuum (Tommerup & Kidby, 1979).

Sterile soil, sand or anhydrous silica gel are usually used as substrates for dehydration.

Soil. The storage of fungi in soil attempts to mimic natural conditions. The method usually involves adding spore or mycelial suspensions in sterile distilled water to twice-autoclaved soil in universal bottles. The soil cultures are incubated at room temperature for up to 10 days until the hyphae have penetrated throughout the soil. The cultures are then stored at 4–7°C with the caps of the bottles loose. The fungus is recovered by sprinkling a few grains of the soil onto a suitable medium followed by incubation under optimal growth conditions (Smith & Onions, 1983). Many fungi survive the procedure and in general survival periods are long (Table 4.5).

Anhydrous silica gel. This preservation method can be used to preserve many fungi that sporulate when grown on artificial media (Table 4.6). Spore suspensions prepared in cooled skimmed milk (4°C) are transferred to cold (−20°C), sterile, non-indicator silica gel in universal bottles. These are incubated in loose-capped bottles at room temperature for 10–14 days until the silica gel crystals separate easily. The caps of the bottles are tightened and the cultures stored at 4–7°C. The fungi can be recovered by sprinkling crystals onto suitable agar media and incubating under optimal growth conditions. Alternatively, an agar plate can be inoculated by drawing silica gel crystals across it and then discarding them. Fungal spores left on the plate are then incubated without any silica gel crystals remaining.

Table 4.5. *Shelf-life of fungi stored in soil*

Fungus	Storage temperature (°C)	Period before transfer (years)
Alternaria spp.	room	5
Aspergillus spp.	room	5
Calonectria spp.	4–7	10–20
Chaetomium spp.	room	5
Circinella spinosa	room	5
Cylindrocarpon spp.	4–7	10–20
Cylindrocladium spp.	4–7	10–20
Fusarium oxysporum	room	5
Fusarium spp.	4–7	10–20
Gibberella spp.	4–7	10–20
Melanospora spp.	4–7	10–20
Nectria spp.	4–7	10–20
Penicillium spp.	room	5
Pseudocercosporella spp.[a]	4	1
Rhizopus nigricans	room	5
Septoria spp.[a]	4	1.75
Thielavia spp.	4–7	15

[a] Host infectivity unimpaired.

Advantages and disadvantages. These two drying methods differ in that preservation in soil allows an initial growth period before desiccation so that both spores and vegetative cells are preserved, whereas silica gel stops growth and metabolism and only fungus spores survive.

Despite successful survival of fungi in soil, Booth (1971a) found that the initial growth period preceding storage may in some cases allow selective growth or abnormal genotypes. In some cases there may be a considerable time-lag before the onset of dormancy which may permit variant vegetative strains to develop and overgrow the wild type, or for saprophytic segregants to overgrow pathogenic ones. In one study 50% of the isolates of *Fusarium acuminatum* that survived storage in soil had been replaced by variant strains, and 67% of the isolates of *F. semitectum* were replaced by non-wild-type strains.

Table 4.6. *Viability of fungi after storage on silica gel for between 8 and 11 years*

	Tested	Viable	% Survival
Mastigomycotina			
Chytridiomycetes	5	0	0
Oomycetes	5	0	0
Zygomycotina			
Zygomycetes	34	20	59
Ascomycotina			
Clavicipitales	1	1	100
Diaporthales	5	3	60
Dothideales	10	4	40
Endomycetales	2	2	100
Eurotiales	4	4	100
Helotiales	1	0	0
Hypocreales	4	3	75
Ophiostomatales	4	1	25
Pezizales	4	3	75
Sordariales	38	30	79
Sphaeriales	6	3	50
Basidiomycotina			
Hymenomycetes	19	10	52
Gasteromycetes	1	0	0
Deuteromycotina			
Coelomycetes	28	21	75
Hyphomycetes	250	202	81

Soil storage has been used successfully to retain viability of fungi such as *Alternaria*, *Pseudocercosporella* and *Septoria* which are not so inherently variable as Fusaria (Table 4.5).

Cultures to be preserved in silica gel must be sporulating (Table 4.6). Representatives of the Mastigomycotina and many species of sporulating Zygomycotina (e.g. *Coemansia, Conidiobolus, Entomophthora, Martensiomyces* and *Piptocephalis*) fail to survive. Only one member of the Mucorales, *Syzigites megalocarpus*, has been found to fail the initial dehydration. Many isolates that cannot be successfully dried will survive liquid nitrogen storage or freeze-drying (Sections 4.3.3 and 4.3.4).

4.4.3 Freeze-drying

Freeze-drying (or lyophilization) is the dehydration of frozen aqueous material through the sublimation of ice under vacuum. Although many sophisticated machines are available (Rowe & Snowman, 1978) freeze-drying does not have to be an expensive technique. Simple reconstruction of the early apparatus used by Raper & Alexander (1945) can bring the costs within reach of most laboratories (Smith & Onions, 1983).

Methods. There are two basic methods of freeze-drying. In one, the suspension is frozen by evaporation under reduced pressure, and in the other the suspension is prefrozen before evacuation. If suspensions of cells are not frozen before evacuation, rapidly escaping water vapour may cause overflow from containers. Centrifugation during vacuum cooling overcomes this problem.

Prefreezing is achieved by several methods. Shelf freeze-driers have refrigerated shelves on which suspensions can be frozen at a controlled rate before drying under vacuum (Fig. 4.3). The vials are sealed or plugged within the chamber, still under vacuum. Suspensions can also be precooled in freezers, in controlled coolers or freezing mixtures prior to evacuation and drying under vacuum. Suspensions for centrifugal freeze-drying are frozen by evaporation in a vacuum chamber. When the suspension is frozen, centrifugation is stopped and drying of the frozen material continues under vacuum. The ampoules are then transferred to a secondary drying machine where drying can be completed over a desiccant such as phosphorus pentoxide before sealing under vacuum.

Suspending medium. Although some fungi can survive freeze-drying on agar disks without a suspending medium (Smith & Onions, 1983), the fungus is usually suspended in a medium to protect it from freezing damage and oxidation. Suspending media commonly used are skimmed milk, serum, peptone, various sugars or mixtures of them. A mixture of skimmed milk (10% v/v) and inositol (5% w/v) is the medium used in several major fungal culture collections.

Fig. 4.3. Shelf freeze-dryer in use at the CAB International Mycological Institute (IMI).

Shelf-life. Freeze-drying was first applied to fungi on a large scale by Raper & Alexander (1945). Ellis & Roberson (1968) later reported successful revival of most of their original cultures. Several surveys have been made of this technique, describing the various methods used and varying degrees of success obtained (Alexander *et al.*, 1980; Smith & Onions, 1983). Most common sporulating fungi can be preserved by freeze-drying (Table 4.7). However some, such as Entomophthorales,

Table 4.7. *Viability of strains preserved by centrifugal freeze-drying and stored at IMI for between 7 and 20 years*

	Tested	Viable	% Survival
Mastigomycotina			
Chytridiomycetes	6	0	0
Oomycetes	29	0	0
Zygomycotina			
Zygomycetes	821	754	92
Ascomycotina			
Ascosphaerales	7	7	100
Clavicipitales	13	8	62
Diaporthales	36	32	89
Diatrypales	2	1	50
Dothideales	236	225	95
Elaphomycetales	1	1	100
Endomycetales	36	36	100
Eurotiales	85	85	100
Gymnoascales	88	86	98
Helotiales	44	42	95
Hypocreales	132	123	93
Microascales	49	49	100
Ophiostomatales	45	44	98
Pezizales	56	51	91
Polystigmatales	13	13	100
Rhytismatales	6	6	100
Sordariales	351	341	97
Sphaeriales	66	44	67
Taphrinales	6	6	100
Basidiomycotina			
Hymenomycetes	92	41	42
Gasteromycetes	5	2	40
Urediniomycetes	3	0	0
Ustilaginomycetes	24	19	79
Deuteromycotina			
Coelomycetes	576	526	91
Hyphomycetes	5091	4822	95

Oomycetes (*Achyla, Pythium, Phytophthora* or *Plectospira*) and hyphal basidiomycetes, are sensitive. Experience has shown that asporogenous strains or fungi with delicate spores fail to survive. Attempts made to freeze-dry sensitive mycorrhizal fungi (Crush & Pattison, 1975) have not been encouraging. Some strains of species such as *Fusarium* (Jong & Davis, 1978) and *Syzigites* (Hwang, 1966) that have been successfully freeze-dried fail to survive, suggesting strain specificity.

Survival and damage in many cases cannot be predicted. To estimate the shelf-life of freeze-dried fungi, accelerated storage tests have been used (Rogan & Terry, 1973). However, time remains the best test for successful storage. Many fungi have been found to survive up to 20 years (Table 4.8) or more (Ellis & Roberson, 1968), but a higher loss of viability and deterioration of properties has been reported after 30 years' storage using some procedures (Jong, Levy & Stevenson, 1984) (Fig. 4.4).

Table 4.8. *Shelf-life of fungi stored after freeze-drying*

Fungus	Method	Inoculum	Storage temperature (°C)	Period before transfer (years)
Absidia spp.	prefrozen	spores	4–8	19
Aspergillus	vacuum	conidia	22–32	20
	cooling		6–7	20
Circinella spp.	prefrozen	spores	4–8	9–19
Cunninghamella spp.	prefrozen	spores	4–8	2–19
Gliocladium spp.	prefrozen	spores	4–8	2–19
Laccaria laccata	prefrozen	basidiospores	−10 or room	0.5
Microsporon canis	vacuum			
	cooling	conidia	6–7	16
Mollerodiscus lentus	prefrozen	apothecia	—	1
Neurospora crassa[a]	prefrozen	ascospores	4	18–20
Penicillium	prefrozen	spores	4–8	2–19
Puccinia[b]	prefrozen	uredospores	2	1
Pyrenochaeta	vacuum			
	cooling	mycelium	10	3
Rhizopus	prefrozen	spores	4	1–19
Trichophyton	vacuum			
mentagrophytes	cooling	conidia	6–7	16
Zygorhynchus				
(12 strains)	prefrozen	spores	4–8	2–19

[a] Retained mutant characteristics.
[b] Retained pathogenicity.

Factors affecting survival. The most critical points of the freeze-drying process are the selection of a suitable suspending medium, optimal cooling rates, maintaining the frozen state during drying, the residual moisture content after drying, and the avoidance of rehydration and contact with oxygen during processing and storage.

The slower cooling rates of centrifugal freeze-drying appear more successful than the more rapid ones resulting from immersion in a freezing mixture (Heckly, 1978). A cooling rate of $1\,°C\,min^{-1}$ has proved best for the cryopreservation of fungi (Hwang, 1966).

To achieve high survival levels, it is important that the temperature of the frozen material should be kept below $-15\,°C$ until the water content is reduced to <5% (Smith, 1986). A number of fungi have shown poor recoveries from freeze-drying because the lowest temperature reached with evaporative cooling in some freeze-drying machines is only $-12\,°C$.

It has been found that the residual moisture of the freeze-dried material must be prevented from falling below 1% (Smith, 1986). Once dried, ampoules can be stored at room temperature. Storage within the temperature range of -70 to $+4\,°C$ may reduce the rate of deterioration that can occur over the long term, but the lower temperatures appear

Fig. 4.4. Lyophilized ampoules stored at the Centraalbureau voor Schimmelcultures (CBS).

unwarranted, as excellent results have been recorded for storage at 4 °C (Heckly, 1978).

Rehydration is an important factor in the recovery of some freeze-dried fungi (Haskins, 1957). The length of time allowed for rehydration, either in broth or distilled water, can be significant (Butterfield, Jong & Alexander, 1974). Rehydration for 24 h in 0.1% peptone has proved to be a useful technique for reviving sensitive fungi, although 30 min rehydration in liquid medium is usually adequate.

Despite the widespread success of freeze-drying, problems have been reported. In rare cases some morphological and physiological character-istics may be lost, sometimes low survival rates suggest the selection of resistant cells, and genetic damage has been observed (Heckly, 1978).

*Advantages and disadvantages.*The advantages of preservation by freeze-drying are that shelf-life is extended and isolates generally retain their morphological, biochemical and physiological characteristics. The total sealing means that there is no risk of contamination or infestation. Ampoules are easily distributed by post and this is an important factor for service collections (Fig. 4.5). However, some isolates fail to survive or have low survival levels, and there is the possibility of selection or genetic damage.

Fig. 4.5. Selection of ampoules and vials containing lyophilized cultures from different culture collections.

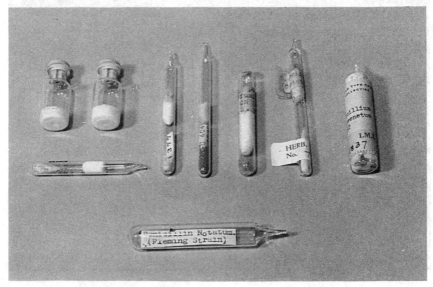

4.4.4 *Freezing*

Lowering the temperature of biological material reduces the rate of metabolism until, when all internal water is frozen, no further biochemical reactions occur and metabolism is suspended. Although little metabolic activity takes place below −70 °C, recrystallisation of ice can occur at temperatures above −130 °C and this can cause damage during storage. Consequently, the storage of microorganisms at the ultra-low temperature of −196 °C in liquid nitrogen is the best preservation method currently available (Fig. 4.6; Table 4.9).

Methods. A variety of cooling rates have been used from 'slow' (1 °C min^{-1}) to 'fast' (200 °C min^{-1}). The final storage temperatures used also vary (−20, −80, −100, −135 and −196 °C), often depending on the facilities that are available. Storage at −196 °C in liquid nitrogen, or in the vapour above it at approximately −150 °C, is superior, as temperatures below −130 °C are necessary for maximum cell stability (see above). Slow cooling rates are more successful and 1 °C min^{-1} is commonly used. The rate of warming also influences viability, but in this case fast rates (*c.* 200 °C min^{-1}) give the highest survival levels (Smith, Morris & Coulson, 1986).

Fig. 4.6. Liquid nitrogen storage tanks in use at the American Type Culture Collection (ATCC).

Table 4.9. *Shelf-life of fungi frozen and stored at sub-zero temperatures*

Species	Growth conditions before freezing	Cryoprotective additives	Cooling (°C min^{-1})	Warming	Recovery	Successful storage period (years) and temp. (°C)
Ascomycotina	spores or mycelium washed from agar surface	glycerol (10% v/v) Me$_2$SO (5% v/v)	1	'rapid'	+	2–18 at −196
Basidiobolus	spores or mycelium washed from agar surface	glycerol (10% v/v) Me$_2$SO (5% v/v)	1	'rapid'	+	1–8 at −196
Basidiomycotina (5 strains)	spores or mycelium washed from agar surface	glycerol (10% v/v) Me$_2$SO (5% v/v)	1	'rapid'	+	1–3 at −196
Deuteromycotina	spores or mycelium washed from agar surface	glycerol (10% v/v) Me$_2$SO (5% v/v)	1	'rapid'	+	2–18 at −196
Entomophthorales	spores or mycelium washed from agar surface	glycerol (10% v/v) Me$_2$SO (5% v/v)	1	'rapid'	+	1–5 at −196
Helminthosporium (12 strains)	spores or mycelium washed from agar surface	None	Immersion in liquid air	'rapid'	80–85%	2 at −189
Mycelia sterila (20 strains)	spores or mycelium washed from agar surface	glycerol 10% v/v) Me$_2$SO (5% v/v)	1	'rapid'	+	0.5–4 at −196
Phytophthora spp.	14–21-day spores or mycelium washed from agar surface	skimmec milk (8.5% w·v) and glycerol (10% v/v)	1	'rapid'	100% germination and infection	2–11 at −196
Pythium spp.	mature cultures washed from agar slopes or mature liquid cultures	glycerol (10% v/v), skimmed milk (8.5% v/v) and glycerol (10% v/v)	1	c. 400 °C min^{-1}	+	2–9 at −196
Saprolegnia spp.	14-day culture mycelium washed from agar surface	glycerol (10% v/v)	1	'rapid'	+	1–6 at −196
Sclerospora spp.	conidia harvested from infected host	Me$_2$SO (10% v/v)	1	'rapid'	10–76%	2 at −196

Storage in mechanical freezers at the higher temperatures is effective for the preservation of many fungi. At domestic deep-freeze temperatures (-17 to $-24\,°C$), many fungi can be stored successfully on agar slopes (Carmichael, 1962). Early methods involved the removal of cultures from the freezer and discarding residual material after thawing and use, but it was found later that if cultures are not allowed to thaw, subcultures can be taken from frozen material and the culture returned to the freezer without loss in viability (Kramer & Mix, 1957). However, some fungi store poorly at these temperatures.

Thawing should be rapid and is usually achieved by immersion of a vial or ampoule in a heated water bath. Once all ice has melted, the culture can be inoculated on to its host or suitable growth medium. Preservation of fungi in polyester film has been developed to eliminate the danger of explosions, increase the storage capacity of refrigerators and facilitate efficient handling and retrieval of cultures (Tuite, 1968). Plastic drinking straws, polypropylene vials or ampoules and wide-necked plastic ampoules can also be used as containers for freezing fungi, or infected plant material for the preservation of obligate plant pathogens.

Other liquid gases can provide temperatures low enough to preserve fungi successfully; for example, liquid air has been used to preserve *Puccinia* urediniospores and the chlamydospores of *Ustilago* (Joshi *et al.*, 1974). There are risks of explosion involved with liquid air and it is considered safer to use liquid nitrogen.

Cryoprotectants. The suspending medium in which the fungi are frozen can be very important, and a cryoprotectant such as dimethyl sulphoxide (Me_2SO) or glycerol is usually used. Glycerol has been used successfully for many fungi (Hwang, 1966). Dimethyl sulphoxide has also been shown to be effective in some cases where glycerol has failed (Hwang & Howells, 1968). Mixtures of cryoprotectants such as Me_2SO and glucose also prove valuable for the protection of fungi (Smith, 1983).

Shelf-life. Freezing and storage in or above liquid nitrogen has proved to be of value for fungi that are difficult to maintain by other means (Table 4.10). For example, mycelial fragments and hyphal tips of *Neurospora crassa* and *Sordaria fimicola* show renewed growth on suitable media after freezing and thawing from liquid nitrogen (Wellman & Walden, 1964). Many fungi retain properties otherwise lost in continuous culture, for example, plant pathogenicity in *Phytophthora* (Dahmen, Staub & Schwinn, 1983).

In general, if a fungus survives freezing, the shelf-life will depend on the storage temperature. An isolate which survives may be successfully stored for 6 months to 2 years at $-20\,°C$, but may survive 20 years or more at storage temperatures below $-130\,°C$.

Factors affecting survival. The initial cooling rate appears to be important and a rate of $1\,°C\ min^{-1}$ has proved to be most successful with fungi.

Storage of frozen material can be in liquid nitrogen or in the vapour

Table 4.10. *Viability of fungi stored in liquid nitrogen at CMI*

	Tested	Viable	% Viability
Mastigomycotina			
Chytridiomycetes	56	9	16
Hyphochytriomycetes	5	3	60
Oomycetes	348	172	50
Zygomycotina			
Zygomycetes	267	254	95
Ascomycotina			
Ascosphaerales	7	6	86
Clavicipitales	15	13	87
Diaporthales	17	17	100
Dothideales	109	97	89
Elaphomycetales	1	0	0
Endomycetales	10	10	100
Eurotiales	56	55	98
Gymnoascales	42	41	98
Helotiales	33	33	100
Hypocreales	39	38	97
Microascales	13	13	100
Ophiostomatales	22	22	100
Pezizales	41	39	95
Polystigmatales	8	6	75
Rhytismatales	7	7	100
Sordariales	178	170	96
Sphaeriales	59	51	86
Taphrinales	5	5	100
Basidiomycotina			
Hymenomycetes	149	143	96
Gasteromycetes	8	8	100
Uredinomycetes	18	10	56
Ustilaginomycetes	6	6	100
Deuteromycotina			
Hyphomycetes	1543	1465	95
Coelomycetes	238	224	94

phase above it. Cracked, faulty or improperly sealed glass ampoules can be dangerous if stored under the liquid. Care must be taken to ensure perfect seals. Many collections use polypropylene cryotubes to avoid the possibility of the explosion of glass ampoules.

Advantages and disadvantages. Cultures of fungi frozen and stored at temperatures below $-130\,^{\circ}\text{C}$ remain viable and stable for variable periods of time (Table 4.9). When stored in sealed ampoules they are free of contamination and the majority of both sporulating and non-sporulating fungi survive well, giving the method a greater range of successful application. The disadvantage of the technique is that apparatus is costly and the supply of liquid nitrogen can prove expensive unless it can be stored in bulk or made on the premises. If the supply of nitrogen fails, then the whole collection can be lost. In hotter climates the evaporation rate of the liquid nitrogen is fast. Furthermore, a regular supply cannot be obtained in some parts of the world and therefore the technique cannot be used. The storage refrigerators must be kept in a well-ventilated room, as constant evaporation of the nitrogen gas could displace the air in the room and suffocate workers. The double-jacketed, vacuum-sealed storage vessels have been known to corrode and rupture. Such a failure may result in the loss of the entire collection.

It is essential that protective clothing is worn and strict safety procedures are followed in all activities involving the use of liquid nitrogen.

4.5   Recommended methods of preservation
4.5.1   *Mastigomycotina*
These fungi are best stored in liquid nitrogen, although some strains do not survive the freezing stages. In these cases storage under mineral oil may be satisfactory for periods of up to 6 months. Alternatively, cultures can be kept viable by storage in water (see Section 4.4.1) and transferred every 2 years. The latter two techniques are not so successful for retaining properties of the fungi, but can be used if stability is not a priority or as a back-up to liquid nitrogen storage. Silica gel cannot be used successfully for this group and at present freeze-dried cultures do not retain viability in the long term.

4.5.2   *Zygomycotina*
Liquid nitrogen storage is recommended for the preservation of Zygomycetes such as *Mucor*, *Rhizopus* and similar genera. Most isolates can be successfully freeze-dried. Not all (*Coemansia*, *Martensiomyces*,

Conidiobolus, Entomophthora, Piptocephalis and *Syzygites*) survive dehydration, particularly on silica gel. Of these genera, only *Piptocephalis* and *Coemansia* strains can be freeze-dried successfully.

4.5.3   Ascomycotina

The majority of the Ascomycetes which can be grown in culture can be freeze-dried and/or stored in liquid nitrogen. Generally, genera and species that do not sporulate well in culture survive poorly. However, most of these have survived long-term storage in liquid nitrogen. Many species do not survive on silica gel.

4.5.4   Basidiomycotina

The higher fungi generally grow only as mycelium in culture and therefore present problems in preservation. Usually such fungi can only be preserved by serial transfer on agar, with or without oil, or stored in liquid nitrogen. The fungi producing thick-walled hyphae can be freeze-dried, but their survival levels are usually low. However, basidiospores harvested from fungi growing in their natural environment can usually be freeze-dried, and will survive other preservation techniques better than the mycelium. Mycelium of wood-inhabiting Basidiomycetes can be grown and maintained on wood chips. The best technique for preservation of all Basidiomycetes is cryopreservation in liquid nitrogen.

4.5.5   Deuteromycotina

Conidial fungi are relatively easy to preserve, freeze-drying being the most widely used technique. Common species such as *Aspergillus, Penicillium* and *Paecilomyces* can be stored for 6 months to 2 years on agar slants in a freezer at −18 °C. Soil storage has been used extensively for the storage of *Fusarium*. Silica gel can be used for the preservation of many conidial fungi, but in tests at CMI 19% of Hyphomycetes studied failed to survive. Liquid nitrogen storage is the most successful preservation technique for this group and is strongly recommended.

In general, freezing and storage in liquid nitrogen is the best preservation technique available for filamentous fungi (Table 4.11). The handling techniques, freezing protocols, cryoprotection and thawing rates can be optimised for particular fungi to obtain maximum survival. Once the fungus has been successfully frozen and stored in liquid nitrogen the storage periods appear to be infinite, since no chemical and very few physical changes can occur at such low temperatures. The problems caused by ice crystal growth during thawing remain, although these are small at rapid rates of warming. The costs of liquid nitrogen and storage equipment make the method more suited to the larger service collections. If strain stability is not critical, smaller collections may prefer to rely on freeze-drying or freezer cabinets, with working cultures held at 4–7 °C.

Table 4.11. *Comparison of methods of preservation*

Method of preservation	Cost		Shelf-life	Genetic stability
	material	labour		
Serial transfer on agar				
(i) Storage at room temperature	low	high	1–6 months	variable
(ii) Storage in the refrigerator	medium[a]	high	6–12 months	variable
(iii) Storage under oil	low	low/medium	1–32 years	poor
(iv) Storage in water	low	low/medium	2–5 years	moderate
(v) Storage in the deep freeze	medium[a]	low/medium	4–5 years	moderate
Drying				
Soil	low	medium	5–20 years	moderate to low
Silica gel	low	medium	5–11 years	good
Freeze-drying	high	initially medium[b]	4–40 years	good
Freezing				
Liquid nitrogen storage	high	low	infinite	good

[a] Refrigerator or deep freeze costs included.
[b] Initial processing is costly depending on the method; subsequent storage is negligible.

5

Identification

D. L. HAWKSWORTH

5.1   Introduction

The numbers of known fungi are vast in comparison with the numbers in other groups of microorganisms used in biotechnological and other industries. Around 64 200 species (including yeasts) are currently known (Hawksworth, Sutton & Ainsworth, 1983), with new species being described at the rate of about 1500 each year. The number being described is limited only by the available mycologists, and the actual number of fungal species in the world may well exceed 250 000.

While only 7000 of these species have yet been grown successfully in pure culture, fungi hitherto known only on a specific host or natural substratum continue to be encountered or cultured for the first time; with appropriate techniques it is clear that many more species could also be cultured. This, together with the appreciation that many of the new species discovered each year are grown in pure culture, is exciting to the biotechnologist in search of strains with significant novel properties.

With such large numbers of fungal species, and a dispersed systematic literature growing at the rate of around 1200 titles each year (*Bibliography of Systematic Mycology* 1943 on), the extent to which a non-specialist can expect to identify isolates to species with confidence is limited. However, some culture collections have specialists in identification on their staff or are associated with scientists able to assist in their identification of isolates and further information on collections providing these services can be obtained by reference to Chapter 1.

Considerable care is needed in identification if confusion is not to be created in the commercial and scientific literature. There are many cases of elegant biochemical, chemical, cytological or ultrastructural studies which have used incorrectly identified material. This can lead to research workers wasting scarce resources searching for the same properties in the wrong strains, or establishing research programmes based on them.

Although culture collections do not always have sufficient numbers of specialists to check the identity of isolates of all groups of fungi at the time of deposit or during preservation procedures, some have access to colleagues with the required specialist knowledge. Nevertheless, it may sometimes happen that cultures are kept under incorrect names applied to them by the original depositor. Time spent authenticating identifications and establishing the purity of strains is a sound investment which may avoid wasting scarce resources and possible future embarrassment. Fortunately, published descriptions and illustrations are available for the more common species (see Section 5.5).

5.2   Characters used in identification

In contrast to bacteria and yeasts, the identification of most groups of filamentous fungi continues to be based almost exclusively on their outward appearance, or morphology, and generally requires the use of a microscope able to achieve a magnification of ×1000. This is a result of both the relatively large numbers of morphological characters exhibited and the considerable number of species known only on their natural host or substratum.

The major divisions within the kingdom Fungi are based on the methods by which the sexual spores are produced (Table 5.1). The Deuteromycotina comprises 'conidial' or 'imperfect fungi' in which no sexual spores are formed; some fungi referred here (especially in the Ascomycotina, see Section 5.5.7) may later be found to have sexual stages, while others may never produce these morphological characteristics. Within the main groups the crucial taxonomic characters are the development and structure of specialised bodies containing the spore-producing organs and the precise nature of the organs themselves. Variations in spore colour, septation and ornamentation are often valuable at the generic level, whereas differences in size tend to be critical at the species level and below.

In identifying fungi it is therefore necessary to know how the spores are produced and the structure of any bodies on which they form or which contain them. Structures containing spores are best examined as freezing-microtome sections mounted in lactophenol-cotton blue (LCB). Mature spores, that is, ones released from their producing organ, should be measured using a calibrated eyepiece micrometer. Spore sizes can vary considerably, even within single strains, so it is always advisable to examine at least 10 mature spores to obtain an accurate measurement. The scanning electron microscope (SEM) is of particular value in the detailed study of fungus spores and is of considerable importance in the identification of *Aspergillus* and *Penicillium* (Fig. 5.1)

In conidial fungi (Deuteromycotina), the gross appearance of colonies developed on agar is of considerable importance in identification. The culture medium, temperature and period of incubation can all affect the colony morphology and so it is important to employ standard pro-cedures and media recommended for particular fungal genera (e.g. CYA for *Penicillium*, see Appendix; Samson & Pitt, 1986).

There is enormous scope for the use of biochemical characters in the identification of filamentous fungi, but this potential is only now starting to be explored in depth. Biochemical tests are not routinely used in identification but many systems used in bacteriology have potential value (Table 5.2) (Bridge, 1985; Bridge & Hawksworth, 1984, 1985) (Figs.

Table 5.1. *The divisions of the kingdom Fungi[a]*

Division	Characteristics	No. of orders	No. of species
Myxomycota[b]	Trophic phase amoebae, an aggregation of amoebae, or a plasmodium	9	625
Mastigomycotina	Thallus unicellular or mycelial; spores and/or gametes motile	8	1160
Zygomycotina	Thallus mycelial and typically non-septate; sexual reproduction resulting in a resting spore; motile cells absent	6	665
Ascomycotina	Thallus mycelial and septate or rarely unicellular; sexual reproduction by ascospores developed inside asci	37	28 650
Basidiomycotina	Thallus mycelial and septate; clamp connections characteristic; sexual reproduction by basidiospores developed from the outside of basidia	23	16 000
Deuteromycotina	Thallus mycelial and septate or unicellular; sexual reproduction generally absent (the parasexual cycle may occur)	N/A	17 000

[a] Based on data included in Hawksworth *et al.* (1983).
[b] This is a separate subkingdom; the other five groups all belong to subking-dom Eumycota.

Table 5.2. *Examples of physiological and biochemical activities used in the characterisation of fungus strains*

Growth in presence of inhibitors (e.g. formalin, preservatives, copper sulphate, phenol, sodium azide, malachite green)

Growth on low-water activity media, at different temperatures and at different pH

Utilisation of carbon sources (e.g. lactose, glucose, soluble starch, sucrose, mannitol, oxalate)

Utilisation of nitrogen sources (e.g. nitrate, nitrite, ammonium, creatine, glycine)

Enzyme activities; hydrolysis (of e.g. aesculin, starch, Tween 80, cellulose, lignin, RNA, casein, gelatin); reduction (of e.g. tetrazolium, tellurite); API ZYM tests (e.g. aryeomidases, chymotrypsinase, galactosidases); isoenzymes (e.g. pectinases)

Secondary metabolite profiles; production of mycotoxins and other compounds, separated and visualised by thin-layer chromatography

---

Fig. 5.1. Scanning electron micrographs (SEM) of *Penicillium* conidia. (A) Echinulate (*P. echinulatum*); (B) lobate (*P. crustosum*); (C) microtuberculate (*P. aurantiovirens*); (D) microverrucate (*P. corylophilum*). Photographs by Dr Z. Lawrence. Crown copyright.

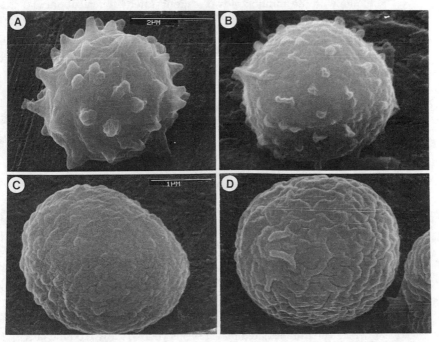

5.2–5.3), as does the use of isoenzyme profiles or 'zymograms' (Cruick-shank & Pitt, 1987). Many studies of such characters have used only a few strains from limited geographical and host, or substratum, ranges, and questions of the reliability and reproducibility of results remain to be answered. The biochemical characterisation of filamentous fungi will be a major asset not only to biotechnologists, but to those concerned with the differentiation of pathogenic from non-pathogenic, or toxic from non-toxic strains of the same fungus species. Such work will also help our understanding of host preferences and substrate utilisation and relate these to the enzymes the cultures produce.

The secondary metabolites of some genera have been studied quite thoroughly and here the presence or absence of particular compounds may be valuable in checking identifications. The development of standardised thin layer chromatographic (TLC) procedures (Frisvad, 1981; Paterson, 1986) will facilitate their routine use.

Fig. 5.2. Screening for enzymic activity. The API ZYM test strip can be used to screen for the presence or absence of 19 specific enzymes in liquids and suspensions. Enzyme activity is shown by a colour change with the addition of two reagents after 4-h incubation. Reproducible results can be obtained for microfungi using conidial suspensions or spent culture fluid.

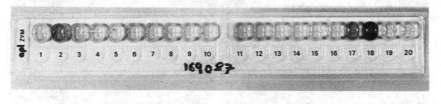

Fig. 5.3. Examples of physiological and biochemical tests. (A) Disc diffusion test for inhibitory substances; conidia are 'seeded' in an agar plate, on which are placed filter paper discs containing known concentrations of suspected inhibitors; after 5–7 days' incubation inhibition is evident as clear zones around the discs. Illustrated is *Penicillium brevicompactum* and discs containing copper sulphate (top left), malachite green (top right), crystal violet (bottom right) and zinc sulphate (bottom left). (B) Testing for lipase activity; fatty acids, such as Tween 80, are incorporated in an agar containing bromocresol purple. The uninoculated plate (left) is yellow; growth on the fatty acid and lipase activity is indicated by a rise in pH and the production of an opaque halo around the colony.

5.3 Reasons for name changes

A common complaint of those concerned with the many different aspects of applied mycology is the apparently endless change of names. This is largely a result of the inadequate level of knowledge of

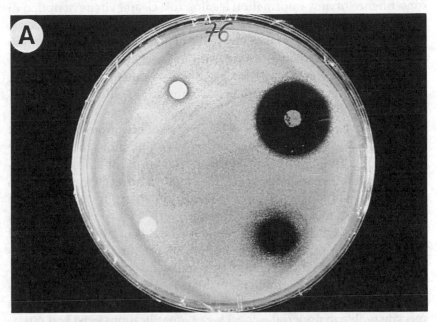

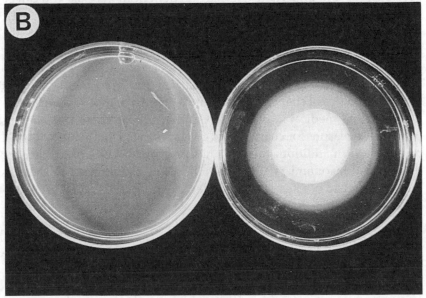

many genera and species, and it mainly arises from a shortage of mycologists with the skills needed to develop more satisfactory taxonomic systems.

Name changes should be welcomed when they arise, since they are a consequence of new information leading to a clearer circumscription or understanding of relationships between particular genera or species. The new names are thus more indicative of genetic relationships and so have a greater predictive value with respect to the properties or behaviour of the remodelled taxa, a development of considerable importance to biotechnologists in search of fungi with particular biochemical attributes.

However, name changes do not always reflect improved levels of knowledge of the fungi themselves, but are due more to inadequate attention to past literature or to misapplications of the rules of nomenclature. While such changes are frustrating and a charge to which mycologists would have to plead guilty, it is perhaps not surprising in view of the fact that there may be as many as 300 000 species names on record. There is a clear need to rectify oversights of predecessors, and the introduction of improved rules of nomenclature is aimed at long-term stability, even though short-term instability may result.

Another reason for name changes is simply the original misidentification of the strain by the depositor. Such mistakes should not occur, as specialists able to undertake checking of identifications exist and can be contacted through the major service culture collections (see Chapter 1).

Nomenclatural problems are recognised internationally. In 1986, the International Commission of the Taxonomy of Fungi (ICTF) of the International Union of Microbiological Societies (IUMS) started to publish (in the IUMS journal, *Microbiological Sciences*) a regular current awareness series of changes in the names of fungi of industrial, biotechnological and medical importance (Cannon, 1986). These publications provide the reasons for changes and guidance as to whether they should be adopted. In addition, the ICTF has prepared a Code of Practice for mycological taxonomists aimed at minimising the number of changes arising from bad practices (Sigler & Hawksworth, 1987) and is establishing a series of subcommissions to promote stability in groups of particular economic importance. The first of these is concerned with *Aspergillus* and *Penicillium*.

5.4   Rules of nomenclature

The nomenclature of all fungi (including yeasts) is governed by the *International Code of Botanical Nomenclature* (ICBN), which is open to modification at each of the International Botanical Congresses held every six years. While it is generally recognised that Fungi are a kingdom distinct from Plantae, this does not preclude their treatment as plants for the purposes of nomenclature. Special provisions for fungi are incorporated into the main body of the Code. The Congress establishes a Special Committee for Fungi and Lichens which advises on all matters of relevance to the fungi.

Microbiologists will find the ICBN a much more complex document to work with than the Bacteriological Code, and the two Codes differ in several important respects. The most significant of these are summarised below, but the ICBN itself should be consulted for definitive information (Voss *et al.*, 1983).

Within the rank of species, fungi can be recognised as belonging to a range of subordinate categories each of which can be given an independent scientific name; these include, in descending order, subspecies, variety, form, and special form.

For names to be nomenclaturally acceptable they must meet criteria for effective and valid publication and be in accordance with the Code (i.e. legitimate). The Code does not rule on taxonomy, placement, position or rank, but only on the correct name to be used when a fungus is considered to belong in a particular position or rank; in cases where the views of mycologists vary, the same fungus may have different nomenclaturally correct names according to the differing opinions. In general, the correct name is the earliest validly published name of the same rank, and for species this is combined with the earliest genus name considered appropriate for the fungus.

In order to be validly published, newly described fungi must now have a diagnosis, or description, in Latin and the nomenclatural type must be indicated. The single element (holotype) designated as the type by the original author as a reference point for the application of the name, must be a permanently preserved dried specimen, dried culture, or slide. Living strains are not accepted as type material, but where possible mycologists are urged to deposit live cultures derived from the holotype with the major service culture collections; these are often loosely referred to as 'types' in culture collecton catalogues but as they have no nomenclatural status are more accurately described as 'ex-type' (i.e. from the nomenclatural type). Attempts to change this ruling on living type material have not found general favour among mycologists as a whole, but the present position can lead to difficulties where the diagnostic features are biochemical or pathological. However, with the development of increasingly sophisticated techniques, this may become less of a problem in the future. Work at the CAB International Mycological Institute has shown that secondary metabolites and DNA are recoverable from dried cultures in the herbarium.

Mycologists have to consider all names published since Linnaeus' *Species Plantarum* of 1753. In 1981, an earlier complex system of starting-point dates for the different groups of fungi was abandoned, but names accepted in certain works (Fries, *Systema mycologicum*, 1821–32; Persoon, *Synopsis fungorum*, 1801) are given priority over earlier names that would otherwise have to be used. Names accepted in these works are said to be 'sanctioned' and this status can be indicated by a colon (:) in author citations when this is considered desirable (e.g. *Botrytis cinerea* Pers.: Fr.).

Special provisions are made for fungi with pleomorphic life cycles (that is, more than one spore type), as it is often desirable to apply different names to the various stages for purposes of precision, and because they may have different biochemical or pathological properties. The name of the sexual phase ('teleomorph') is the one to be used for a fungus in all its stages, but any asexual phase reproducing by conidia ('anamorph') can be given an independent scientific name which only applies to that phase. The detailed rules on this provision were extensively modified in 1981. The short-term inconvenience caused was recognised, but considered a modest price to pay for a simpler system. For example, the name *Penicillium brefeldianum* B. Dodge when introduced in 1933 included both teleomorph and anamorph phases; the teleomorph, however, belongs to *Eupenicillium* and in 1976 was correctly called *Eupenicillium brefeldianum* (B. Dodge) Stolk & Scott as Dodge's name is typified by the teleomorph part of the included material. A separate name was required for the anamorph, and *P. dodgei* Pitt was introduced for this in 1980. Anamorph names never take priority over names typified by material of the teleomorph.

Mycologists rarely use the rank of subspecies (subsp.), but variety (var.) is not infrequently employed. For plant pathogens 'special form' (f.sp.) is used for races only able to attack particular host plants, but which cannot be separated on morphological criteria.

There is at present no equivalent to the bacterial list of approved names for fungi, but there exists a procedure known as 'conservation', designed to ensure the maintenance of well-known generic names which would otherwise have to be changed by a strict application of the ICBN. Such cases are reviewed and voted on by the Special Committee for Fungi which makes recommendations to the next International

Botanical Congress. This procedure has recently been extended to the names of species of major economic importance. In addition, a list is maintained of rejected species names that cannot be used because of the confusion this would cause.

5.5 Literature references for identification

5.5.1 *General*

Ainsworth, G. C., Sparrow, F. K. & Sussman, A. S. (eds.) (1973). *The Fungi. An Advanced Treatise.* New York & London: Academic Press. Vols. IV A–B. 621+504 pp. (Includes keys to most accepted genera and extensive literature lists; the standard multi-authored reference work.)

Arx, J. A. von (1981). *The Genera of Fungi Sporulating in Pure Culture*, 3rd edn. Vaduz: J. Cramer. 424 pp.

Domsch, K. H., Gams, W. & Anderson, T.-H. (1980). *Compendium of Soil Fungi.* London: Academic Press. 2 vols., 859+405 pp.

Ellis, M. B. & Ellis, J. P. (1985). *Microfungi on Land Plants. An Identification Handbook.* London & Sydney: Croom Helm. ix+818 pp.

Gams, W., van der Aa, H. A., van der Plaats-Niterink, A. J., Samson, R. A. & Stalpers, J. A. (1987). *CBS Course in Mycology*, 3rd edn. Baarn: Centraalbureau voor Schimmelcultures.

Hawksworth, D. L. (1974). *Mycologists' Handbook.* Kew: Commonwealth Mycological Institute. 231 pp.

Hawksworth, D. L., Sutton, B. C. & Ainsworth, G. A. (1983). *Ainsworth & Bisby's Dictionary of the Fungi*, 7th edn. Kew: Commonwealth Mycological Institute. xii+445 pp.

Holden, M. (1982). Guide to the literature for the identification of British fungi, 4th edn. Part 1, General, Myxomycota, Mastigomycotina, Zygomycotina, Ascomycotina. *Bulletin of the British Mycological Society* **16**, 36–55. Part 2, Basidiomycotina, Deuteromycotina. *Bulletin of the British Mycological Society* **16**, 92–112.

Kirk, P. M. & Hawksworth, D. L. (1986). Aspects of the nomenclature of fungi. In *Biological Nomenclature Today*, ed. W. D. L. Ride & T. Younès, pp. 54–8. IUBS Monograph Series No. 2. Oxford: IRL Press.

Sims, R. W., Freeman, P. & Hawksworth, D. L. (1988). *Key Works to the Fauna and Flora of the British Isles and Northern Europe.* Oxford: Clarendon Press. viii+312 pp.

Stevens, R. B. (1974). *Mycology Guidebook.* Seattle & London: University of Washington Press. xxiv+703 pp.

5.5.2 *Habitat keys and lists*

Animals and man

Emmons, C. W., Binford, C. H., Utz, J. P. & Kwon-Chung, K. J. (1977). *Medical Mycology.* 3rd edn. Philadelphia Lea & Febiger. ix+592 pp.

Howard, D. H. (ed.) (1983–5). *Fungi Pathogenic for Humans and Animals.* 3 vols. New York & Basel: Marcel Dekker. 652, 553, 381 pp.

McGinnis, M. R. (1980). *Laboratory Handbook of Medical Mycology*. New York: Academic Press. xiii+661 pp.

Entomogenous fungi

Samson, R. A., Evans, H. C. & Latge, J. P. (1988). *An Atlas of Entomopathogenic Fungi*. Berlin: Springer.

Food spoilage fungi

Pitt, J. I. & Hocking, A. (1985). *Fungi and Food Spoilage*. Sydney: Academic Press. 413 pp.

Samson, R. A., Hoekstra, E. S. & van Oorschot, C. A. N. (1984). *Introduction to Food-borne Fungi*, 2nd edn. Baarn: Centraalbureau voor Schimmelcultures. 248 pp.

Industrial fungi

Onions, A. H. S., Allsopp, D. & Eggins, H. O. W. (1981). *Smith's Introduction to Industrial Mycology*, 7th edn. London: Edward Arnold. viii+398 pp.

Marine fungi

Johnson, T. W. & Sparrow, F. K. (1961). *Fungi in Oceans and Estuaries*. Weinheim: J. Cramer. xxii+668 pp.

Kohlmeyer, J. & Kohlmeyer, E. (1979). *Marine Mycology. The Higher Fungi*. New York: Academic Press. xiv+690 pp.

Plant pathogenic fungi

Arx, J. A. von (1987). *Plant Pathogenic Fungi*. Beihefte zur Nova Hedwigia no. 87. Berlin & Stuttgart: J. Cramer. 288 pp.

Ellis, M. B. & Ellis, J. P. (1985). *Microfungi on Land Plants. An Identification Handbook*. London & Sydney: Croom Helm. ix+818 pp.

CAB International Mycological Institute (1964 on). *Descriptions of Plant Pathogenic Fungi and Bacteria*. Kew.

Soil fungi

Barron, G. L. (1986). *The Genera of Hyphomycetes from Soil*. Baltimore: Williams & Wilkins. Reprint 1972. Huntington, New York: R. E. Krieger Publishing. 346 pp.

Domsch, K. H., Gams, W. & Anderson, T. -H. (1980). *Compendium of Soil Fungi*. London: Academic Press. 2 vols., 859+405 pp.

Thermophilic fungi

Cooney, D. G. & Emerson, R. (1964). *Thermophilic Fungi*. San Francisco & London: W. H. Freeman. xii+188 pp.

5.5.3 Myxomycetes (Myxogastres)

Farr, M. L. (1981). *How to Know the True Slime Molds*. Dubuque, Iowa: Wm. C. Brown Co. 132 pp., 159 figs.

Martin, G. W. & Alexopoulos, C. J. (1969). *The Myxomycetes*. Ames, Iowa: University of Iowa Press. 561 pp., 41 col. pls.

Martin, G. W., Alexopoulos, C. J. & Farr, M. L. (1983). *The Genera of Myxomycetes*. Ames, Iowa: University of Iowa Press. 102 pp., 41 col. pls.

5.5.4 Protosteliomycetes (Protostelia)

Olive, L. S. (1975). *The Mycetozoans*. New York & London: Academic Press. x+293 pp.

5.5.5 Dictyosteliomycetes (Dictyostelia) and Acrasiomycetes (Acrasea)

Raper, K. B. (1984). *The Dictyostelids*. Princeton University Press. x+453 pp.

5.5.6 Mastigomycotina and Zygomycotina

Fuller, M. S. (ed.) (1978). *Lower Fungi in the Laboratory*. Athens, Georgia: University of Georgia. 212 pp., many illustrations.

Sparrow, F. K. (1960). *Aquatic Phycomycetes*, 2nd edn. Ann Arbor: University of Michigan Press. 1187 pp.

Oomycetes

Plaats-Niterink, A. van der (1981). Monograph of the genus *Pythium*. *Studies in Mycology* **21**, 1–242.

Waterhouse, G. M. (1968). The genus *Pythium* Pringsheim. *Mycological Papers* **110**, 1–203.

Waterhouse, G. M. (1970). The genus *Phytophthora* de Bary, 2nd edn. *Mycological Papers* **122**, 1–59.

Zygomycetes

Benjamin, R. K. (1959–65). The merosporangiferous Mucorales. *Aliso* **4**, 321–433; **5**, 11–19, 273–322; **6**, 1–10. Reprinted as one volume 1967. Lehre: J. Cramer.

Hesseltine, C. W. (1955). Genera of Mucorales with notes on their synonymy. *Mycologia* **47**, 344–63.

O'Donnel, K. L. (1979). *Zygomycetes in Culture*. Athens, Georgia: University of Georgia. 257 pp., many illustrations.

Schipper, M. A. A. & Stalpers, J. A. (1984). A revision of the genus *Rhizopus*. *Studies in Mycology* **25**, 1–34.

5.5.7 Ascomycotina (Ascomycetes)

General

Arx, J. A. von & Müller, E. (1954). Die Gattungen der amerosporen Pyrenomyceten. *Beitrage zur Kryptogamenflora der Schweiz* **11** (1), 1–434.

Breitenbach, J. & Kränzlin, F. (1981). *Pilze der Schweiz*. Vol. 1. *Ascomycetes*. Lucerne: Edition Mykologia. 310 pp.

Cannon, P. F., Hawksworth, D. L. & Sherwood-Pike, M. A. (1985). *The British Ascomycotina. An Annotated Checklist*. Kew: CAB International Mycological Institute. 302 pp. (Includes numerous references to literature.)

Dennis, R. W. G. (1978). *British Ascomycetes*, 3rd edn. Lehre: J. Cramer. xxxii+456 pp.

Eriksson, O. & Hawksworth, D. L. (1987). Outline of the ascomycetes – 1987. *Systema Ascomycetum* **6**, 259–337.

Müller, E. & von Arx, J. A. (1982). Gattungen der didymosporen Pyrenomyceten. *Beitrage der Kryptogamenflora Schweiz* **11** (2), 1–922.

Dothideales

Arx, J. A. von & Müller, E. (1975). A re-evaluation of the bitunicate ascomycetes with keys to families and genera. *Studies in Mycology* **9**, 1–159.

Eriksson, O. (1981). The families of bitunicate ascomycetes. *Opera Botanica* **60**, 1–220.

Sivanesan, A. (1984). *The Bitunicate Ascomycetes and their Anamorphs*. Vaduz: J. Cramer. 701 pp.

Eurotiales

See also under Hyphomycetes.

Pitt, J. I. (1980) ('1979'). *The Genus Penicillium and its Teleomorphic States Eupenicillium and Talaromyces*. London: Academic Press. 634 pp.

Raper, K. B. & Fennell, D. I. (1965). *The Genus Aspergillus*. Baltimore: Williams & Wilkins. 686 pp.

Samson, R. A. (1979). A compilation of the Aspergilli described since 1965. *Studies in Mycology* **18**, 1–38.

Samson, R. A. & Pitt, J. I. (eds.) (1986) ('1985'). *Advances in Penicillium and Aspergillus Systematics*. NATO Advanced Study Institute Series A 102. New York & London: Plenum Press. 483 pp.

Stolk, A. C. & Samson, R. A. (1972). The genus *Talaromyces*. *Studies in Mycology* **2**, 1–65.

Stolk, A. C. & Samson, R. A. (1983). The ascomycete genus *Eupenicillium* and related *Penicillium* anamorphs. *Studies in Mycology* **23**, 1–149.

Onygenales

Benny, G. L. & Kimbrough, J. W. (1980). A synopsis of the orders and families of Plectomycetes with a key to genera. *Mycotaxon* **12**, 1–91.

Currah, R. S. (1985). Taxonomy of the Onygenales. Arthrodermataceae, Gymnoascaceae, Myxotrichaceae and Onygenaceae. *Mycotaxon* **24**, 1–216.

Ophiostomatales
Olchowecki, A. & Reid, J. (1974). Taxonomy of the genus *Ceratocystis* in
   Manitoba. *Canadian Journal of Botany* **52**, 1675–1711. (Key 70 species.)

Sordariales (Coronophorales)
Arx, J. A. von (1975). On *Thielavia* and some similar genera of ascomycetes.
   Studies in Mycology **8**, 1–28.
Arx, J. A. von, Guarro, J. & Figueras, M. J. (1986). The ascomycete genus
   Chaetomium. Beihefte zur Nova Hedwigia **84**, 1–162.
Lundqvist, N. (1972). Nordic Sordariaceae *s. lat. Symbolae botanicae upsalienses*
   20 (1), 1–374.

5.8   *Basidiomycotina (Basidiomycetes)*

General
Heim, R. (1969). *Les Champignons d'Europe*, 2nd edn. Paris: Boubée & Cie.
   680 pp.
Michael, E. & Hennig, B. (1958–70). *Handbuch für Pilzfreunde.* Jena: G.
   Fischer. 5 vols., many col. pls.

Agaricales
Henderson, D. M., Orton, P. D. & Watling, R. (1968). *British Fungus Flora:
   Agarics and Boleti. Introduction.* Edinburgh: Her Majesty's Stationery Office.
   58 pp.
Kühner, R. & Romagnesi, H. (1953). *Flore analytique des Champignons
   superieurs.* Paris: Mason & Cie. 557 pp.
Singer, R. (1975). *The Agaricales in Modern Taxonomy*, 3rd edn. Vaduz: J.
   Cramer. 912 pp.
Watling, R. & Watling, A. E. (1980). *A Literature Guide for Identifying
   Mushrooms.* Eureka, California: Mad River Press. 121 pp.

Aphyllophorales
Eriksson, J. & Ryvarden, L. (1973–84). *The Corticiaceae of North Europe.* Oslo:
   Fungiflora. Vols. 2–7 so far published.
Jülich, W. & Stalpers, J. A. (1980). *The Resupinate Non-poroid Aphyllophorales
   of the Temperate Northern Hemisphere.* Amsterdam: North-Holland
   Publishing Company.
Stalpers, J. A. (1978). Identification of wood-inhabiting fungi in pure
   culture. *Studies in Mycology* **16**, 1–248.

5.9   *Deuteromycotina (Fungi Imperfecti)*

Coelomycetes
Morgan-Jones, G., Nag Raj. T. R. & Kendrick, W. B. (1972–5). *Icones
   Generum coelomycetum I–VII.* University of Waterloo Biology Series.
Sutton, B. C. (1980). *The Coelomycetes. Fungi Imperfecti with pycnidia, acervuli
   and stromata.* Kew: CAB International Mycological Institute. 696 pp., 397 figs.

Hyphomycetes

Barron, G. L. (1968). *The Genera of Hyphomycetes from Soil.* Baltimore: Williams & Wilkins. Reprint 1972. Huntington, New York: Krieger Publishing. 346 pp.

Booth, C. (1971). *The Genus Fusarium.* Kew: CAB International Mycological Institute. 237 pp.

Burgess, L. W. & Liddell, C. M. (1983). *Laboratory Manual for Fusarium Research.* University of Sydney. 162 pp.

Carmichael, J. W., Kendrick, W. B., Conners, I. C. & Sigler, L. (1980). *Genera of Hyphomycetes.* Edmonton: University of Alberta Press. 386 pp.

Ellis, M. B. (1971). *Dematiaceous Hyphomycetes.* Kew: CAB International Mycological Institute. 608 pp., 419 figs.

Ellis, M. B. (1976). *More Dematiaceous Hyphomycetes.* Kew: CAB International Mycological Institute. 507 pp., 383 figs.

Gams, W. (1971). *Cephalosporium-artige Schimmelpilze (Hyphomycetes).* Stuttgart: G. Fischer. x+262 pp.

Gerlach, W. & Nirenberg, H. (1982). The genus *Fusarium* – a pictorial atlas. *Mitteilungen aus der Biologischen Bundesanstalt für Land- und Forstwirtschaft, Berlin-Dahlem* **209**, 1–406.

Nag Raj, T. R. & Kendrick, B. (1976). *A Monograph of Chalara and Allied Genera.* Waterloo: Wilfrid Laurier University Press.

Nelson, P. E., Toussoun, T. A. & Cook, R. J. (eds.) (1982). *Fusarium: Diseases, Biology, and Taxonomy.* University Park & London: Pennsylvania State University Press. 457 pp.

Pitt, J. I. (1980) ('1979'). *The Genus Penicillium and its Teleomorphic States Eupenicillium and Talaromyces.* London: Academic Press. 634 pp.

Pitt, J. I. (1985). *A Laboratory Guide to Common Penicillium Species.* North Ryde, NSW: CSIRO Division of Food Research. 183 pp.

Raper, K. & Fennell, D. I. (1965). *The Genus Aspergillus.* Baltimore: Williams & Wilkins. 686 pp.

Samson, R. A. (1979). A compilation of the Aspergilli described since 1965. *Studies in Mycology* **18**, 1–38.

Samson, R. A. (1974). *Paecilomyces* and some allied Hyphomycetes. *Studies in Mycology* **6**, 1–119.

Samson, R. A. & Pitt, J. I. (eds.) (1986) ('1985'). *Advances in Penicillium and Aspergillus Systematics.* NATO Advanced Study Institute Series A 102. New York & London: Plenum Press. 483 pp.

Seifert, K. A. (1985). A monograph of *Stilbella* and some allied Hyphomycetes. *Studies in Mycology* **27**, 1–235.

Subramanian, C. V. (1972). *Hyphomycetes.* New Delhi: Indian Council of Agricultural Research. 930 pp.

6

Patent protection for biotechnological inventions

I. J. BOUSFIELD

6.1 Introduction

This chapter is intended to give the reader who is unfamiliar with patents an introduction to the patent system as it applies to biotechnology, and a general guide to the procedures and pitfalls involved in obtaining patent protection for biotechnological inventions. For a detailed discussion of the whole subject of patents in biotechnology and a review of the variety of national patent systems the reader is referred to the excellent texts by Crespi (1982), Beier, Crespi & Straus (1985) and Straus (1985). It is not possible here to provide a step-by-step guide to getting a patent in every country in the world, for, despite an overall similarity, variations between different national patent laws are manifold, and professional help is necessary to guide even the experienced inventor through their complexities. The present account does no more than skim the surface of what is a complex and often fascinating subject; for this reason a short list of selected publications which illustrate in more detail many of the points raised here is given in Section 6.6, Further reading.

6.2 Basis of the patent system
6.2.1 Principles

The principle (if not the practice) of the patent system is straightforward: the inventor of a new product or process publicly discloses the details of his invention and in return he is granted for a limited period a legally enforceable right to exclude others from exploiting it. In this way the inventor's ingenuity is acknowledged and rewarded, while at the same time further technical progress is encouraged by the public dissemination of information about the invention.

6.2.2   *Criteria for patentability*

To qualify for patent protection, an invention must meet the following major criteria.

Novelty. The invention must be *new*. Most countries apply the test of 'absolute novelty', which means that if prior knowledge of it exists anywhere in the world (not merely in the country where patent protection is sought) then the invention belongs to the state of the art ('prior art') and is not patentable. The prior art is held in these countries to include anything the inventor himself may have said or written about his invention. Exceptions to this rule of absolute novelty are the patent systems of the USA and Canada, where publications by the inventor made not more than one (USA) or two (Canada) years before a patent application is filed in that country do not destroy novelty. 'Grace periods' of six months are also allowed by Australia, New Zealand and Japan, but only in respect of certain kinds of publications made by the inventor, for example at certain scientific meetings. Under nearly all patent systems, publications made after the date that the patent application is filed (the 'priority date') do not jeopardise protection for that particular subject matter in that particular application. It must be remembered, however, that they will form part of the prior art against which any *future* applications will be assessed.

Inventiveness. The invention must show evidence of an *inventive step*, that is it must not be 'obvious' from the state of the art to anyone 'skilled in the art'. In simple terms this means that the average expert in the field under consideration could not reasonably have predicted the invention as an obvious or logical outcome of what he already knew.

Utility (industrial applicability). The invention must have a *practical use*, which in the USA means exactly that, but in nearly all other patent systems means that it must also be capable of industrial application. Most countries, however, hold that medical methods for the direct treatment of the human or animal body are not susceptible to industrial application and are therefore not patentable. The major exception to this is again the USA.

Disclosure. The details of the invention must be *disclosed* by means of a patent specification (see Section 6.5.2 below) so that the invention is described in sufficient detail to allow a skilled person to reproduce it. This is normally done by means of a written description supplemented where necessary by drawings. However, in one major category of biotechnological inventions – those involving the use of new microorganisms and certain other novel living materials – a written description is not usually considered sufficient for the purposes of disclosure. In such cases it is argued that no matter how carefully the description may be worded, if the microorganism itself is not available, the invention cannot be reproduced. Therefore, many countries require the deposit of new microorganisms in a recognised culture collection to ensure their public availability. This unique aspect of biotechnological patent procedure is dealt with in some detail later in this chapter.

Exclusions from patentability. As well as meeting the criteria listed above, an invention must not be of a kind which is excluded from patentability by its very nature. Exclusions relating specifically to biotechnological inventions are discussed later, but in general terms patents cannot be obtained for mere discoveries, theories, computer programs, literary works, musical compositions, aesthetic creations and illegal or offensive devices.

6.3 Kinds of biotechnological inventions

There are four main kinds of biotechnological invention: products, compositions, processes, and use or methods of use (Crespi, 1982).

6.3.1 *Products*

These inventions are exactly what the word suggests – tangible new materials or entities. They include organisms themselves (e.g. bacteria, fungi), parts of organisms (e.g. cell lines), substances produced by either of these (e.g. enzymes, antibiotics), and substances obtained by or employed in recombinant DNA techniques (e.g. plasmids, DNA molecules).

Product inventions can be the subject of two broad kinds of patent claim: the 'product *per se*', where patent protection is sought for the product itself, regardless of the method of manufacture, and the more limited 'product-by-process', where protection is sought for the product obtained by a particular process.

6.3.2 *Compositions*

These inventions are mixtures of substances or organisms, the individual components of which may already be known, but which in combination can be shown to display a new property or exert a new effect.

6.3.3   *Processes*

These inventions are methods for the manufacture of products, and include bioconversions, fermentations, and methods of isolation, purification or cultivation. Some process inventions are genuinely new methods for obtaining novel or known products, but others are known methods applied in new situations or used in the manufacture of novel products.

6.3.4   *Use and methods of use*

Methods of use include processing or treating materials (e.g. industrial raw materials or agricultural products), non-medical treatments of humans or animals, 'off the body' medical methods (e.g. a method of diagnosis carried out on a sample taken from a patient), methods of testing (e.g. quality control) and in a few countries, notably the USA, medical treatment of humans and animals. Also, the new medical use of a substance previously unknown to have that use is protectable in European Patent Convention (EPC) countries.

6.4   Patentability of biotechnological inventions

6.4.1   *Inventions involving new plants and animals*

Plant varieties. By far the most common form of legal protection for new plant varieties is the plant variety right (although in the USA special 'plant patents' are available for asexually propagated plants). Several countries are now party to the International Convention for the Protection of New Varieties of Plants (UPOV) which aims to harmonise national practices as far as possible (International Convention, 1978). In these countries, plants that are protected by plant variety rights are usually specifically excluded from patentability. Plant variety rights in general are intended to allow the commercial plant breeder a monopoly on the production of propagating material for the purposes of commercial marketing, its offer for sale and its marketing. Plant variety rights are easier to obtain than is patent protection as there is no requirement for inventiveness or for reproducible disclosure, but they are more limited in scope in that neither the plant itself nor consumables produced from it (e.g. fruit for eating, grain for milling) are protected.

Plant variety rights were introduced essentially to cover varieties developed by traditional breeding methods, and it is such varieties that are excluded from patentability in many countries. Thus Article 53(b) of the EPC, to which 13 countries belong (Table 6.1) states the following:

> European patents shall not be granted in respect of . . . plant or animal varieties or essentially biological processes for the production of plants or animals; this provision does not apply to microbiological processes or the products thereof.

The same exclusion is found where the national laws of individual European countries have been harmonised with the EPC and is also contained in the laws of several other countries, e.g. the German Democratic Republic, Mexico, Sri Lanka, Thailand and Yugoslavia. Plant and animal varieties, but not essentially biological processes, are also excluded from patentability in China.

The exclusion of plant varieties from patent protection is a contentious issue. Straus (1985) has pointed out that current systems for the protection of plant varieties were introduced when plant breeding methods did not permit the breeder to fulfil the normal criteria of patentability. However, the advent of new technologies, particularly genetic manipulation techniques, for the production of new plant varieties has meant that these requirements can now be met. Therefore, there seems to be a good argument in favour of allowing the developer of such varieties the right to choose between protection under the patent system or through plant variety rights (Beier *et al.*, 1985; Straus, 1985).

Table 6.1. *Countries party to the European Patent Convention at 1 January 1987*

Austria
Belgium
France
Germany (Federal Republic)
Greece
Italy
Liechtenstein
Luxembourg
Netherlands
Spain
Sweden
Switzerland
United Kingdom

Animal varieties. Although a special form of protection for new animal varieties (which are not, however, regarded as inventions) is available in the Soviet Union, there is in general no special system of legislation for their protection. Most of the countries that exclude plant varieties from patentability also exclude animal varieties, and in those countries that do not, the position is not altogether clear. However, recent court decisions in the USA (*in re* Diamond & Chakrabarty, 1980) and Canada (*in re* Abitibi, 1982) suggest that animals may be patentable provided that the requirements for enabling disclosure, that is repeatability, are met (Straus, 1985).

Processes for the production of animals and plants. Many patent laws, including those of the EPC, specifically deny patent protection for 'essentially biological processes' for the production of plants or animals; microbiological processes, however, are not included in this provision. This terminology may not be entirely clear and perhaps needs some explanation. Stated simply, an essentially biological process is considered to be one in which the result is achieved with a minimum of human technical intervention. The example given by the European Patent Office (EPO) in its guidelines to examiners is a method for selectively breeding horses in which human intervention is limited to bringing together animals having particular characteristics. On the other hand, a process for treating a plant to promote or suppress its growth, e.g. a method of pruning or of applying stimulatory or inhibitory substances, would not be considered to be essentially biological, since although a biological process is involved, the essence of the invention is technical.

Given this definition of 'essentially biological' and the exemption of microbiological processes from it, the normal criteria for patentability can be applied to methods for producing plants by, for instance, genetic manipulation involving the use of vectors in microbial hosts, or by somatic cell hybridisation. However, matters are less certain under the EPC in respect of some processes for the production of new animal varieties, even though such processes may meet the test of not being essentially biological. This is because a further exclusion from patentability is found in Article 52(4) of the EPC, which states the following:

> Methods for treatment of the human or animal body by surgery or therapy and diagnostic methods practised on the human or animal body shall not be regarded as inventions which are

susceptible of industrial application . . . [See also Utility in Section 6.2.2 above.]

Straus (1985) has expressed the fear that some present and future approaches to animal breeding, such as techniques of embryo transfer, could be denied patent protection by this provision. In support of his argument he cites a recent decision by the UK Comptroller of Patents, in which an application involving just such a technique was rejected as being a method of treatment by surgery.

In contrast to the European system, the patent laws of the USA, Japan and China do not exclude essentially biological processes. Furthermore, the US laws do not exclude methods for treating animals (or humans); therefore the problems presented by the European system in respect of patenting processes involved in animal breeding do not exist in the USA.

Tissue cultures. Animal cell lines and plant tissue cultures (and, in Japan, seeds) are generally considered to be in the same category as microorganisms for patent purposes. Thus they are subject to the provisions applied to microbiological inventions as discussed below. As regards plant cells, however, the US Patent Office makes a distinction between undifferentiated cell lines used, for instance, to produce a particular substance, and cells which are capable of differentiation and which are used simply to reproduce the whole plant.

6.4.2   *Inventions involving microorganisms*

Applied microbiology in its broadest sense is a major facet of modern biotechnology, and any discussion of patents in biotechnology inevitably must focus on the peculiar problems posed by microbiological inventions. In fact so great has been the attention given to these problems in patent circles that the patent legislation of an increasing number of countries contains specific provisions for inventions involving the use of microorganisms, and one international convention (the Budapest Treaty; see below) deals entirely with microorganisms (Budapest Treaty and Regulations, 1981).

It should be said that the term 'microorganism' is used in patent circles in a very loose sense and encompasses entities that strictly speaking are not microorganisms, e.g. cell lines and plasmids. Indeed, the word is intentionally not defined in the Budapest Treaty so as to avoid undue constraints being imposed upon the application of the Treaty, and in the words of the World Intellectual Property Organisation (WIPO) commentary on the draft Treaty (WIPO, 1980), it 'need not correspond to usage in some scientific circles'. Unfortunately, the use of such inexact terminology has led to uncertainty in some quarters as to what is or is not a microorganism. Because of this, the present author [acting on behalf of the World Federation for Culture Collections (WFCC) Patents Committee] has proposed to WIPO that the expression 'living material' be used instead of the word microorganism, particularly in regard to the Budapest Treaty. The word 'living' was defined as 'that material which under appropriate conditions is able to replicate itself, or which at least possesses the functional genes necessary to direct its own replication'. This definition has two advantages: first, it avoids insoluble philosophical arguments about where chemical reactivity ends and life begins, and second, it excludes non-living biological materials such as enzymes. In the present chapter, although the word microorganism is used for ease of reference, it should be taken to mean living material as defined above.

Microbiological inventions may be found in all of the categories of biotechnological invention outlined in Section 6.3 above. In general, microbiological processes and the (inanimate) products obtained by them can be considered analogous to chemical processes and products, and obtaining patent protection for them in most countries is nowadays fairly straightforward, provided that the basic criteria for patentability are fulfilled. Less straightforward, however, is the patenting of microorganisms as products either *per se* or as products-by-process. It is these inventions above all others that demonstrate the difficulties of determining the borderline between 'discoveries' and 'inventions', of what is 'new' and what is not, and of ensuring sufficiency of disclosure.

Naturally occurring microorganisms. A previously unknown naturally occurring microorganism that is left in its natural state is universally regarded as a discovery and is unpatentable. However, the degree of human intervention considered necessary to turn such a discovery into a patentable invention (assuming it has a practical use) varies between different countries. The extent of this variation was demonstrated by the official replies to a questionnaire on patent protection in biotechnology distributed to governments in 1982 by the Organization for Economic Cooperation and Development (OECD). These responses were reviewed in detail by Crespi (1985). In those countries that do permit naturally occurring organisms to be patented, isolation and purification of the organism are general prerequisites, after which various constraints are applied, mainly relating to novelty and unexpected properties. Thus, for example, the UK and the EPO require an organism to be 'new' in the sense of being hitherto *unknown*, whereas in Canada a new organism is one that does not already *exist* in nature [in this connection, Crespi (1985) has commented pointedly on the illogicality of equating 'unknown' with 'not previously existing']; the Federal Republic of Germany requires that 'certain changes occur during isolation so that the isolated microorganism is not identical with that occurring in nature'; Denmark requires naturally occurring organisms to have unforeseen properties. The USA permits the patenting of naturally occurring organisms as 'biologically pure cultures'.

Non-naturally occurring microorganisms. After the much-publicised Chakrabarty case in the USA in 1980, in which a genetically manipulated strain of *Pseudomonas* was held not to be a product of nature but a human invention patentable *per se*, there are unlikely to be any unusual problems in obtaining patent protection for 'artificial' microorganisms (bacterial recombinants, hybridomas, etc.), other than in countries which do not permit the patenting of any kinds of microorganism.

Sufficiency of disclosure. As already mentioned (Disclosure, in Section 6.2.2 above), one of the fundamental requirements of the patent system is that the details of an invention must be disclosed in a manner sufficient to allow a skilled person to reproduce the invention. Microbiological inventions present particular problems of disclosure in that more often than not repeatability cannot be ensured by means of a written description alone. In the case of an organism isolated from soil, for instance, and perhaps 'improved' by mutation and further selection, it would be virtually impossible to describe the strain and its selection sufficiently to guarantee another person obtaining the same strain from soil himself. In such a case, the strain itself forms an essential part of the disclosure. In view of this an increasing number of countries require a 'new' microorganism (i.e. one not already generally available to the public) to be deposited in a recognised culture collection whence it can subsequently be made available at some stage in the patent procedure. As a general principle, an invention should be reproducible from its description at the time that the patent application is filed. In the case of an invention involving a new organism, therefore, most patent offices require a culture of the organism to be deposited not later than the filing date of the application (or the priority date if priority is claimed from an earlier application – see Section 6.5.3 below). Exactly when a strain becomes available varies according to the patent laws of different countries, and is a much debated question dealt with in more detail below (Release of samples). Since an invention must also be reproducible throughout the life of the patent, a microorganism deposited for patent purposes must remain available for at least this length of time. Most countries provide for a considerable safety margin in this respect, and availability for at least 30 years is a common requirement.

The Budapest Treaty. In order to obviate the need for inventors to deposit their organism in a culture collection in every country in which they intend to seek patent protection, the 'Budapest Treaty on the International Recognition of the Deposit of Microorganisms for the Purposes of Patent Procedure' was concluded in 1977 and came into force towards the end of 1980 (Budapest Treaty and Regulations, 1981). Under the Budapest Treaty certain culture collections are recognised as 'International Depositary Authorities' (IDAs), and a single deposit made in any one of them is acceptable by each country party to the Treaty as meeting the deposit requirements of its own national laws. Any culture collection can become an IDA provided that it has been formally nominated by a contracting state, which must also provide assurances that the collection can comply with the requirements of the Treaty. At 1 January 1987 there were 14 IDAs and they and the kinds of organisms they accept are listed in Table 6.5.

The Budapest Treaty provides an internationally uniform system of deposit and lays down the procedures which depositor and depository must follow (see Section 6.5.5 below), the duration of deposit (at least 30 years or 5 years after the most recent request for a culture, whichever is later), and the mechanisms for the release of samples. The Treaty does not, however, concern itself with the *timing* of deposit nor, in the main, of release; these are determined by the relevant national laws. Likewise, the recipients of samples (other than patent offices and people with the depositor's authorisation) are referred to merely as 'parties legally entitled': exactly *who* such parties are and under what conditions they may obtain samples are again determined by national law. Twenty-one states and the European Patent Office (EPO) are now party to the Budapest Treaty and are listed in Table 6.2.

National deposit requirements. The requirements of various countries for the deposit and release of microorganisms for patent purposes are summarised in Table 6.3. Deposit is a statutory requirement under Rule 28 of the EPC and under the national laws of several of its member countries. In those EPC countries not having a statutory provision under their national law, deposit is such an established requirement of patent offices that it amounts to the same thing for all practical purposes. Most of these countries follow EPC practice in requiring the deposit to be made by the filing or priority date. An exception to this is the Netherlands, where deposit is required before the second publication (see next section) of the patent application. Most other European countries do not have specific requirements as yet, but nevertheless advise that deposits should be made, usually along the lines of the EPC.

In many countries outside Europe, deposit is an established or recommended practice, and some patent offices (in Japan, USA, USSR, for example) have specific requirements for deposit. In almost all cases deposit must be made by the filing or priority date. In the USA, however, as a consequence of a recent court decision (*in re* Lundak, 1985), deposit may in certain circumstances be made after filing but before the issuance of a US patent.

All countries party to the Budapest Treaty (Table 6.2) must recognise a deposit made in an IDA but not all *require* deposits to be made in IDAs. Thus, for example, France, Germany, Switzerland, the UK, the USA and the EPC will recognise other culture collections that can comply with their particular requirements. Hungary accepts deposits made in collections on its own soil, but the only deposits it will recognise

Table 6.2. *Countries party to the Budapest Treaty at 31 July 1987*

Australia	Liechtenstein
Austria	Netherlands
Belgium	Norway
Bulgaria	Philippines
Denmark	Spain
Finland	Sweden
France	Switzerland
Germany (Federal Republic)	UK
Italy	USA
Hungary	USSR
Japan	European Patent Office (EPO)[a]

[a] The EPO is not, strictly speaking, a party to the Treaty, since it is not a country but an intergovernmental organisation. Article 9 of the Treaty provides for such organisations to file a declaration stating that they accept the obligations and provisions of the Treaty. The EPO has filed such a declaration.

Table 6.3. *National requirements (mandatory or recommended) for deposit and release of microorganisms*

Country	Deposit by	Earliest release	Earliest general availability[a]	Restrictions on distribution and use of samples	Minimum storage period (years)
Australia	F/P	1st pub.	1st pub.	as for UK	30
Austria	F/P	–	–	–	–
Belgium	F/P	1st pub.	grant	as for EPO	30
Bulgaria	F/P	grant	grant	as for UK	–
Canada	F/P	grant	grant	none	life of patent
Denmark	F/P	1st pub.	grant	as for EPO	30
Finland	F/P	1st pub.	grant	as for EPO	30
France	F/P	1st pub.	grant	as for EPO	30
Germany	F/P	1st pub.	1st pub.	until patent expires, sample must not be passed to 3rd parties or outside purview of German law	20
Hungary	F/P	1st pub.	1st pub.	sample must not be passed to 3rd party	20
Ireland	F/P	–	–	–	–
Italy	F/P	?	?	as for EPO	?
Japan	F/P	2nd pub.	2nd pub.	sample must not be passed to 3rd parties until patent expires and must be used only for research purposes	life of patent
Liechtenstein	2nd pub.			as for Switzerland	
Netherlands	F/P	2nd pub.	2nd pub.	none	life of patent
New Zealand	F/P	–	–	as for EPO	–
Norway	F/P	1st pub.	grant	as for EPO	30
Portugal	F/P	–	–	–	–
Spain	F/P	1st pub.	1st pub.	–	–
Sweden	F/P	1st pub.	grant	as for EPO	30
Switzerland	F/P	?	?	sample must not be passed to 3rd parties	30

UK	F/P	1st pub.		sample must not be passed to 3rd parties until patent expires and must be used only for experimental purposes	30
USA[b]	F/P	grant		none	30
USSR	–	–		–	–
EPO	F/P	1st pub.	grant	if applicant chooses, available only to independent expert before grant; must not be passed to 3rd parties before patent expires and must be used only for experimental purposes	30

F/P, filing or priority date as applicable; pub., publication of application; –, no provisions or provisions not known; ?, conflicting information from different sources.

[a] General availability means sample publicly available at least in country where application has been filed.

[b] The Lundak decision (1985) may mean that in certain cases deposit may be made later.

elsewhere are those made in IDAs. The Japanese patent office, however, will recognise deposits outside Japan only if they have been made under the Budapest Treaty or have been 'converted' to Budapest Treaty deposits (see 'Converted' deposits in Section 6.5.5 below), regardless of their previous public availability. It must be remembered that deposits made under the Budapest Treaty can only be made in IDAs.

Most countries not party to the Budapest Treaty accept deposits made in any internationally known culture collection which will comply with their requirements; in some cases the collection is required to furnish a declaration as to the permanence and availability of the deposit.

Release of samples. Microorganisms deposited to comply with requirements for disclosure must become available to the public at some stage of the patenting procedure. Unlike a written description, however, the microorganism is the physical essence of the invention itself and because of this the exact conditions of release are a matter of great concern to patent practitioners. There are as yet no internationally uniform release conditions, but three main kinds of system operate at present.

In the USA, patent applications are not published until the patent is granted, and a microorganism deposited for patent purposes need not be made available until then. From the date of grant, the organism must be publicly available without any restriction. The major advantage of this system to the inventor is that his microorganism does not have to be released until he has an enforceable right. Furthermore, if he is *not* granted a patent then his microorganism need never become available. Thus under the US system an inventor is never put in the position of having to allow access to his organism when he has no legal protection.

In Japan and the Netherlands patent applications are published twice, first 18 months after the filing date or priority date and before the application has been examined (Section 6.5.4 below), and second (for the purposes of opposition by third parties) when the application has been accepted. A microorganism deposited in connection with the application must be made available at the date of second publication, i.e. once the patent office has decided to grant a patent. Thus under the Japanese and Dutch systems the inventor again has an enforceable right at the time he is required to make his organism available. In Japan he is afforded a further measure of protection in that recipients of cultures must not pass them on to third parties and must use them only for experimental purposes. This provision does not apply under Dutch law, however.

A dual publication system is also operated by the EPC and by the countries party to it. However, in contrast to the Japanese requirements, a microorganism deposited in connection with an EPC application must be made available at the date of *first* publication, i.e. before any enforceable right exists. This practice reflects the prevailing philosophy of European patent authorities that the organism is regarded as an integral part of the disclosure and therefore should become available at the same time as the written description. Originally cultures had to be available to anyone requesting them, subject to the recipient giving certain undertakings of rather doubtful value, but in response to pressure from users of the system the appropriate rule (Rule 28) of the EPC was amended to provide more protection for the inventor. Rule 28(4) now permits the applicant to opt to restrict the availability of his organism at first publication to an independent expert acting on behalf of a third party. The expert, who is chosen by the requesting party from a list held by the EPO, is not permitted to pass cultures of the strain to anyone else. After second publication, the strain becomes generally available, but at this stage the inventor has an enforceable right.

Although this so-called 'expert solution' applies in respect of applications filed with the EPO itself, it is at present part of the national law of only a minority of member countries of the EPC (France, Italy, Sweden). None, however, permits recipients of cultures to pass them on to third parties.

Need for deposit. In principle most countries require deposit only when repeatability of the invention cannot be ensured without it. Thus, for example, it should not be necessary to deposit a new recombinant strain if the procedure for constructing the novel plasmid and transforming it into a host can be described in sufficient detail to allow an expert to produce the same recombinant for himself (given, of course, that the original vector and host are already generally available). In practice, however, applicants in such cases sometimes choose to deposit in order to avoid all risk of their application being rejected on the grounds of insufficient disclosure. Some applicants, on the other hand, prefer to take this risk.

In cases where the microorganism is already generally available from a culture collection, the situation is perhaps more straightforward. Some countries (e.g. Germany, USA, USSR) require the applicant to furnish a declaration signed by the culture collection and stating that the organism in question is in fact available and will remain available for the period dictated by the relevant national law. In this connection, it is worth noting that the USA in some cases presently requires such a declaration – at least for deposits made outside the USA – even where an organism has been deposited under the Budapest Treaty. As mentioned earlier, the Japanese patent office will recognise the availability of strains from culture collections outside Japan only if they have been deposited under the Budapest Treaty.

6.5   Practical considerations

So far this chapter has been concerned with the principles of biotechnological patents and the requirements of various countries; now the essentially practical aspects of the patenting process must be considered. To use specific examples, either actual or hypothetical, for this purpose would give too narrow a picture. Therefore, the more general question of seeking patent protection for an unspecified invention involving the use of a new (i.e. not already generally available) microorganism will be considered.

6.5.1   *The patent agent (patent attorney)*

The job of the patent agent is, put simply, to obtain a patent on behalf of the applicant. The agent's knowledge and understanding of patent procedures world-wide are essential to guide the applicant through the complex business of seeking patent protection, helping to draft the technical description, formulating the claims, dealing with the patent authorities, ensuring deadlines are met and so on. However, his technical knowledge of the invention and its background cannot be expected to equal that of the inventor, who must therefore be prepared to spend time and effort in familiarising him with every aspect. In his turn, the agent can often offer valuable advice on areas where more experimental work might be done before filing in order to make the application as strong as possible. The importance of these considerations is shown by the fact that many large firms have full-time patent agents on their staff to ensure that their inventions are adequately protected. The small organisation and the academic inventor, therefore, are well advised to obtain the services of a professionally qualified patent agent if they are considering applying for patent protection.

6.5.2 *Disclosing the invention*

Premature disclosure. As mentioned earlier, making a full disclosure of an invention is the applicant's side of the bargain that will give him a legal monopoly, and is a fundamental prerequisite for obtaining a patent. However, the disclosure must be made in the proper way and at the right time. Above all the invention must not be disclosed prematurely, for its novelty (see Section 6.2.2 above) will be assessed by most patent offices in the light of what is already known (the state of the art) on the day the patent application is filed. The prevailing state of the art includes any contributions the applicant himself may have made to it, whether orally, by visual display, or by display or sale of a product. Thus to avoid premature disclosure, the academic inventor in particular must abjure the normal practices of discussing his findings with other workers or publishing them in scientific journals until he has filed his patent application. Information revealed by breach of the applicant's confidence does not jeopardise a patent application, but since breach of confidence is often difficult to prove, the wisest course is to make no disclosure until the application has been safely filed. These strictures do not wholly apply (the exact conditions vary) in relation to those few countries, notably the USA, that allow a 'grace period' (see Section 6.2.2 above). However, even here it must be remembered that although the relevant national law may allow disclosure during a grace period preceding the basic national filing, such a disclosure will be prejudicial to subsequent foreign filings.

The patent specification – technical description. The patent specification contains the written disclosure or technical description of the invention and the patent claims, which state the scope and kinds of monopoly being asked for. The precise wording of the patent specification is of great importance and it is here that the skill of the patent agent comes into play in ensuring that the description (supplemented by deposit – see Section 6.5.5 below) fulfils the requirements of disclosure and that the claims are drafted to afford the best protection.

The technical description is exactly that; the invention is described in detail in scientific and technical terms – it is not addressed to the layman – and put into the context of the field to which it applies, the problems it aims to solve and the way in which solutions are achieved. The preferred format of the description is well established and typically includes the following: field, background, object and summary of the invention, followed by the detailed description of the invention. Crespi (1982) has discussed the layout of the technical description more fully, with actual examples, and the reader is referred to his book for more information. As far as the present account is concerned, the first point is that the description must clearly convey the novelty, inventiveness and industrial applicability of the invention, and must describe the methodology in sufficient detail (worked examples being usual, but not mandatory) to enable a skilled person to reproduce the invention for himself and show that it works in accordance with the claims of the inventor. Second, the technical description must also describe any new microorganism involved in the invention. Clearly the accession number assigned to the organism by the culture collection in which it has been deposited must be quoted, but beyond that the extent of characterisation required varies between countries. The most extensive requirements are those of the Japanese patent office, which gives in its 'examination standard' a detailed list of the properties which should be recorded. The EPO has less stringent guidelines, and at the other extreme the Netherlands will accept deposit of the organism in lieu of any characterisation. Many countries expect the kind of taxonomic data that would be used in scientific publications, although most do not insist on it and accept deposit as a means of offsetting deficiencies in the written characterisation. In general, an applicant is well advised to provide characterisation data 'to the extent available to him' (Crespi, 1985).

The patent specification – claims. The claims are perhaps the most important part of the patent specification as far as the applicant is concerned, because they set out precisely the extent of the protection being sought. This is particularly so in, for example, the UK and USA where great attention is paid to the exact wording of the claims. Any loophole left here can leave the inventor exposed to competition from which he might otherwise have been protected. In Germany, on the other hand, the claims are viewed less literally in that more attention is given to them as indicators of the basic inventive idea. The EPO adopts a middle course, trying to strike a balance between the rights of the inventor and those of third parties. Again, the reader is referred to Crespi (1982) and to Ruffles (1986) for more detailed discussion about patent claims; only the salient points will be given here.

Normal practice is to make the scope of the initial claims as broad as possible, leaving it to the patent office to object if it believes that too much is being claimed for the invention. In the final event, the applicant may feel that the degree of protection he has been allowed is less than it ought to be, but this is better than finding that he is the author of his own misfortune by having claimed too little in the first place. Nevertheless, the claims must not be extravagant; they must be based on the description and be supported by it. Thus the greater the degree of novelty and ingenuity indicated by the description, and the wider the variety of worked examples given, then the broader the claims that are likely to be accepted.

Claims are usually presented as a numbered set. The first is often the broadest and is the general claim; this is followed by subclaims, each defining by example particular aspects of the general claim, and each generally being narower in scope than the one before it. The subclaims represent fall-back positions if the broader claims are held invalid. Claims of more than one kind should be included wherever possible, e.g. a new chemical compound, a microbiological process of producing it, the organism used in the process claimed *per se*, a method of diagnosis using the new compound, and a kit incorporating the new compound and conventional reagents for the diagnosis. Table 6.4 gives examples of sets of patent claims relating to different cell types.

6.5.3   *Filing the application*

A single application filed in one country will result in patent protection only in that country. Therefore when the disclosure of an invention is likely to lead to serious foreign competition the normal course is to seek protection in several countries. Fortunately this does not have to be done all at once and the first application is usually filed in the applicant's own country. The patent office gives this application a number and, more importantly, a filing date. The significance of this first filing date ('priority date') is that it establishes the priority of the invention; in other words any later applications made in that country by other people for the same invention are pre-empted by it. Furthermore, provided that the applicant files any corresponding foreign applications within 12 months of his basic national filing, the original priority date is also recognised by nearly all overseas countries. However, the same priority date cannot be claimed for material not included in the basic national application (the 'priority document').

The EPO also recognises the original priority date for applications filed with it within 12 months of the basic national filing. The advantage of the European system is that an application filed with the EPO results

Table 6.4. *Examples of sets of patent claims*

US Patent no. 4,567,146

We claim:

1. A recombinant plasmid characterised in that it contains DNA of (1) a first Rhizobium plasmid identifiable as being the same as the plasmid pVW5JI or pVW3JI of lower molecular weight present in the culture of the strain of *Rhizobium leguminosarum* NCIB 11685 or 11683 respectively and (2) a second Rhizobium plasmid found in bacteria of another strain of *Rhizobium leguminosarum*, said second plasmid having Rhizobium genes coding for nodulation, nitrogen fixation and hydrogen uptake ability but which is non-transmissible.

2. A method of preparing a culture of bacteria of the genus Rhizobium, which method is characterised in that

(1) in a first cross, a donor strain of Rhizobium, containing (a) a Rhizobium plasmid lacking genes coding for nodulation but which is transmissible, is crossed with a recipient strain of Rhizobium, carrying (b) a Rhizobium plasmid having Rhizobium genes coding for nodulation, nitrogen fixation and hydrogen uptake ability but which is non-transmissible, whereby a transconjugant strain carrying a plasmid which is formed from said plasmids (a) and (b) and is a conjugal precursor of a recombinant plasmid (c) having genes coding for nodulation, nitrogen fixation and hydrogen uptake ability and being transmissible is obtained;

(2) said transconjugant strain is separated from donor and recipient strains and cultured to produce a substantially pure culture thereof;

(3) in a second cross, the transconjugant strain from the first cross is used as a donor strain and crossed with a plasmid-containing recipient strain whereby a transconjugant strain carrying a recombinant plasmid (c) is obtained; and

(4) said transconjugant strain from the second cross is separated from donor and recipient strains and cultured to produce a substantially pure culture thereof.

3. A method according to claim 2 characterized in that the transmissible plasmid (a) carries at least one drug-resistance gene.

4. A method according to claim 3 characterized in that the transmissible plasmid is pVW5JI or pVW3JI, identifiable as being the same as the plasmid of lower molecular weight present in the culture of a strain of *Rhizobium leguminosarum* NCIB 11685 (pVW5JI) or NCIB 11683 (pVW3JI), and a kanamycin-resistant transconjugant strain is separated in each cross.

5. A method according to claim 2 characterized in that the transmissible plasmid (a) contains a selectable determinant.

6. A method according to claim 2, characterized in that the donor and recipient strain are of the species *Rhizobium leguminosarum*.

7. A method of impairing hydrogen uptake ability to bacteria of the genus Rhizobium, which method is characterized in that (1) a strain of *Rhizobium leguminosarum* NCIB 11684 or NCIB 11682, as a donor strain, is crossed with a recipient strain of *Rhizobium leguminosarum* to produce a kanamycin-resistant transconjugant strain, said recipient strain being one which permits selection of the transconjugant strain against the donor and recipient strains and which allows the transconjugant strain to be selected against when used as a donor in a subsequent cross with another strain of *Rhizobium leguminosarum*, (2) said

Table 6.4. (*cont.*)

transconjugant strain is separated from the donor and recipient strains and cultured to produce a substantially pure culture thereof; (3) in a second cross the transconjugant strain obtained from the first cross is used as a donor strain and crossed with a recipient strain of *Rhizobium leguminosarum* to produce a kanamycin-resistant transconjugant strain and (4) said transconjugant strain from the second cross is separated from the donor and recipient strains to produce a biologically pure culture thereof.

8. A method according to claim 7 characterized in that the recipient strain for the first cross is auxotrophic and has resistance to a drug other than kanamycin.

9. A method according to claim 7 or 8 wherein the recipient strain for the second cross is a naturally occurring strain.

10. A Rhizobium plasmid pIJ1008 having Rhizobium genes coding for streptomycin and kanamycin resistance, nodulation, nitrogen fixation and hydrogen uptake properties, which is transmissible and which is the plasmid of lowest molecular weight present in the culture of a strain of *Rhizobium leguminosarum* NCIB 11684 by virtue of the fact that it migrates the fastest on agarose gel in a gel electrophoresis determination in which a gel of 0.7% agarose in Tris-borate buffer of pH. 8.3 is subjected to electrophoresis at 25 mA and 100 volts at 4° C for 16 to 20 hours in the dark.

11. A Rhizobium plasmid pIJ1007 having Rhizobium genes coding for streptomycin and kanamycin resistance, nodulation, nitrogen fixation and hydrogen uptake properties, which is transmissible and which is the plasmid of lowest molecular weight present in the culture of a strain of *Rhizobium leguminosarum* NCIB 11682 by virtue of the fact that it migrates the fastest on agarose gel in a gel electrophoresis determination in which a gel of 0.7% agarose in Tri-borate buffer of pH 8.3 is subjected to electrophoresis at 25 mA and 100 volts at 4° C for 16 to 20 hours in the dark.

12. A biologically pure culture of bacteria of the genus Rhizobium characterized in that it contains a plasmid selected from the group consisting of pIJ1008 and pIJ1007.

13. A culture according to claim 12 of bacteria of the species *Rhizobium leguminosarum*.

14. A biologically pure culture of bacteria of the genus Rhizobium containing a recombinant plasmid characterized in that said plasmid contains DNA of (1) a first Rhizobium plasmid identifiable as being the same as the plasmid pVW5JI or lower molecular weight present in the culture of the strain *Rhizobium leguminosarum* NCIB 11685 or 11683 respectively and (2) a second Rhizobium plasmid found in bacteria of another strain of *Rhizobium leguminosarum*, said second plasmid having Rhizobium genes coding for nodulation, nitrogen fixation and hydrogen uptake ability but which is non-transmissible.

US Patent no. 4,546,082
   What is claimed is:
   1. A DNA expression vector capable of expressing in yeast cells a product which is secreted from said yeast cells, said vector comprising at least a segment of alpha-factor precursor gene and at least one segment encoding a polypeptide.
   2. A DNA expression vector according to claim 1 wherein said segment encoding a polypeptide is an insertion into said alpha-factor precursor gene.
   3. A DNA expression vector according to claim 1 wherein said segment

Table 6.4. (*cont.*)

encoding a polypeptide is a fusion at a terminus of said alpha-factor precursor gene.

4. A DNA expression vector according to claims 2 or 3 wherein coding sequences for mature alpha-factor are absent from said segment of alpha-factor precursor.

5. A DNA expression vector according to claim 1 wherein said polypeptide is somatostatin.

6. A DNA expression vector according to claim 1 wherein said polypeptide is ACTH.

7. A DNA expression vector according to claim 1 wherein said polypeptide is an enkephalin.

8. A yeast strain transformed with a DNA expression vector of claim 1.

9. A method for producing a DNA expression vector containing alpha-factor gene comprising the steps of
   (a) transforming a <u>MAT</u> alpha <u>2 leu 2</u> yeast strain with a gene bank constructed in plasmid YEp13;
   (b) selecting for leu transformants from the population formed in step (a);
   (c) replacing the transformants from step (b) and
   (d) screening for alpha-factor producing colonies.

10. A DNA expression vector formed according to the method of claim 9.

UK Patent no. 1,346,051

What we claim is:

1. *Fusarium graminearum* Schwabe deposited with the Commonwealth Mycological Institute and assigned the number I.M.I. 145425 and variants and mutants thereof.

2. *Fusarium graminearum* Schwabe I-7 deposited with the Commonwealth Mycological Institute and assigned the number I.M.I. 154209.

3. *Fusarium graminearum* Schwabe I-8 deposited with the Commonwealth Mycological Institute and assigned the number I.M.I. 154211.

4. *Fusarium graminearum* Schwabe I-9 deposited with the Commonwealth Mycological Institute and assigned the number I.M.I. 154212.

5. *Fusarium graminearum* Schwabe I-15 deposited with the Commonwealth Mycological Institute and assigned the number I.M.I. 154213.

6. *Fusarium graminearum* Schwabe I-16 deposited with the Commonwealth Mycological Institute and assigned the number I.M.I. 154210.

7. Fungal cultures containing a strain of *Fusarium graminearum* Schwabe I.M.I. 145425 or a mutant or variant thereof in a culture medium in which this strain is present in a culture medium containing or being supplied with nutrients or additives necessary for the sustenance and multiplication of the strain, the medium having a pH between 3.5 and 7 and the temperature of the medium being maintained at a precise value within the range of between 25 and 34° C.

8. A method for cultivating a strain of *Fusarium graminearum* Schwabe I.M.I. 145425 or a mutant or variant thereof wherein the strain is present in a culture medium containing or being supplied with nutrients or additives necessary for the sustenance and multiplication of the strain, the medium having a pH between 3.5 and 7 and the temperature of the medium being maintained at a precise value within the range of between 25 and 34° C.

9. A method for the preparation of variants of *Fusarium graminearum* Schwabe

Table 6.4. (cont.)

I.M.I. 145425 which comprises growing the parent strain I.M.I. 145425 under
continuous culture conditions with carbon limitation in a fermentation.

10. A method for the preparation of variants of *Fusarium graminearum*
Schwabe I.M.I. 145425 which comprises growing the parent strain I.M.I. 145425
on a glucose based medium at 25° to 30° C under continuous culture conditions
at a dilution rate of 0.10 to 0.15 hrs. $^{-1}$ with carbon limitation in a fermentation for
1100 hours.

11. A method as claimed in claim 10 wherein the resulting proliferated
variants are isolated by dilution plating.

12. A method for the preparation of variants of *Fusarium graminearum*
Schwabe I.M.I. 145425 substantially as described with reference to Examples 1 to
5 hereinbefore set forth.

13. Fungal cultures containing *Fusarium graminearum* Schwabe I.M.I. 145425
or mutants or variants thereof substantially as described with reference to any
one of Examples 6 to 11 hereinbefore set forth.

UK Patent no. 1,300,391
What we claim is:

1. A human embryo liver cell line having the characteristics of cells deposited
with the American Type Culture Collection under number CL99.

2. A cell culture system comprising cells derived from the human embryo
liver cell line designated by A.T.C.C. number CL99 in a nutrient culture medium
therefore.

3. A virus culture system comprising cells derived from the human embryo
liver cell line designated by A.T.C.C number CL99 inoculated with a virus
capable of replication in said cells, and a nutrient culture medium adapted to
support growth of the virus-cell system.

4. A culture system according to claim 2 or 3, wherin the nutrient culture
medium contains Eagle's minimum essential medium and heat-inactivated
foetal calf serum.

5. A culture system according to claim 4, wherein the nutrient culture
medium contains Eagle's minimum essential medium, heat-inactivated foetal
calf serum, sodium bicarbonate and one or more antibiotics.

6. A virus cultivation process which comprises maintaining a viable culture of
cells derived from the human embryo liver cell line designated by A.T.C.C.
number CL99 in a nutrient culture medium, inoculating the culture with a virus
to which the cells are susceptible and cultivating the virus in the culture.

7. A process according to claim 6, wherein the virus is of the group consisting
of adenoviruses, San Carlos viruses, ECHO viruses, arthropodborne group A
viruses and other arboviruses, pox viruses, myxoviruses, paramyxoviruses,
picornaviruses, herpes viruses and the AR 17 haemovirus.

8. A process according to claim 6, wherein the virus is a hepatitis virus.

9. A process according to claim 7, wherein the virus is of the group consisting
of adenovirus types 2, 3, 4, 5, 7 and 17, San Carlos virus types 3, 6, 8 and 49,
ECHO virus type 11, Sindbis virus, vaccinia virus, influenza A2 virus and other
influenza viruses, Sabin poliovirus type 1 and other poliomyelitis viruses, and
the AR—17 haemovirus.

10. A virus whenever cultivated by the process of any of claims 6 to 9.

11. Antigenic material obtained from a virus according to claim 10.

Table 6.4. (*cont.*)

---

12. A vaccine comprising a virus according to claim 10, in an administrable form and dosage.
13. A vaccine comprising antigenic material according to claim 11, in an administrable form and dosage.
14. A vaccine comprising antibodies produced by a virus according to claim 10 or antigenic material according to claim 11, in an administrable form and dosage.
15. A cell culture system substantially as described in Example 3.
16. A virus cultivation process substantially as described in Example 3.

---

in a clutch of 'national' patents, valid in those countries party to the EPC that the applicant has designated as being territories in which he wants patent protection. The applicant must, however, designate these countries at the outset; he can drop some of them later by not paying renewal fees, but he cannot add to them. Use of the EPC system is not mandatory in Europe, however; if an inventor wishes, he can instead file separate national applications in individual countries. In fact, if protection is not required in more than two or three European countries, the national route may be cheaper.

The first steps along the road to obtaining patent protection involve drafting the patent specification, filing the basic national application and, within 12 months, filing the appropriate foreign applications (redrafting and developing further the original specification if appropriate). As Crespi (1982) has pointed out, a year is not long when account is taken of the need to evaluate the importance of the invention, decide on the extent of foreign patenting desired and implement the decision. Implementation involves drafting the final specification, sending documents around the world and possibly having the specification translated into other languages. In this last respect it is all too easy, say, for the English-speaking applicant to forget that the German Patent Office will expect an application to be written in German. (When filing with the EPO, however, the application may be drafted in English, French or German.) Thus although the patent agent will take care of all the documentary procedures, consulting the applicant where necessary, the latter cannot afford to relax completely at this stage. Attention must also be paid, of course, to ensuring that by the filing date (or, where applicable, the priority date) the microorganism used in the invention has been deposited in a suitable culture collection. This will be discussed in considerable detail later (see Section 6.5.5 below); for the present it is more convenient to follow the progress of the patent application itself.

6.5.4   *Patent office procedures*

Once the final national and/or foreign applications have been filed and the microorganism deposited, the applicant must wait for his application to be processed by the patent office. The exact procedures and the time they take vary widely between countries. Therefore, only a broad outline, illustrated with a few examples, can be given here.

All major patent offices carry out a novelty search, which is usually a patent and literature search, followed by a critical or substantive examination of the application. Under the EPC and many European national systems, the search and examination are treated separately. After the search, a report is sent to the patent agent, pointing out material (including earlier patents) considered relevant to the application. Under the EPC, this report and the patent application itself should be published 18 months after the priority (basic national filing) date, although in practice, delays in the issuance of the search report are common. At this point, the claims can be modified by the applicant if it seems unlikely that they will be accepted as they stand. Also with the publication of the application, the deposited microorganism becomes available (to varying extents) under most European systems (see Release of samples in Section 6.4.2 above). For the application to proceed further, the applicant must now ask for a substantive examination to be made of it. This request must be made within a specified period or the application will lapse.

Under the Japanese system, patent applications are published after 18 months, but there is no search or examination unless and until the applicant requests it, which he must do within seven years. In Japan and the USA, both the search and substantive examination are carried out before a report is issued to the applicant.

In all systems, a written response to the patent examiner's report must be made within a certain period or the application may fail by default. There usually then follows a variable period of negotiation with the examiner ('prosecution of the application') as to how broad the final claims should be in view of the prior art. If agreement is reached, then the application is accepted. If not, it is refused and the applicant and his agent must then consider whether to pursue an appeal to a higher tribunal.

Negotiations with the patent authorities and the meeting of deadlines are usually taken care of by the patent agent. Unless questions of a highly technical nature are raised, the involvement of the inventor himself in these proceedings is generally minimal.

After an application has been accepted, in most countries the patent is then granted and published (for the first time in the USA and Canada). At this point the deposited microorganism becomes available for the first time in the USA, Canada, Japan and (unless the EPC route has been followed) the Netherlands. Under the EPC system the microorganism, available since the first publication only to an independent expert if the applicant has so opted, becomes generally available.

Major patent offices permit a period immediately after the patent application has been allowed or the patent has been granted for it to be challenged by third parties. The extent of this 'opposition' period varies considerably between countries. Thus, for instance, the EPC allows a nine month period for after-grant revocation of a European patent (before it becomes a collection of national patents). Japan allows an opposition period of three months before grant. In the USA there is no opposition procedure as such. Instead, anyone can ask for a re-examination of any patent, regardless of when it was granted, provided that he can cite pertinent prior art previously unconsidered by the Patent Office and which is sufficient to convince the Office that the issue should be re-opened. Exceptionally, a UK patent can be revoked by application to the Patent Office at any time after grant.

The length of time for which a patent lasts once it has been granted also varies between different countries. In the USA and Canada, for instance, the period is 17 years from the date of grant, regardless of how long the application has been pending the outcome of negotiations between the applicant and the patent office. In Europe, on the other hand, the term is 20 years from the application date. In Japan, it is 15 years from publication for opposition purposes, or 20 years from the application date, whichever is the shorter. Maintenance of the patent for its full term is subject to periodic renewal fees, non-payment of which will result in the patent lapsing.

6.5.5   Depositing the microorganism

With an invention involving the use of a new microorganism, that is, one not already available to the public, there is one other vital act to be performed at an early stage of the patenting procedure. The microorganism must be deposited in a suitable culture collection in order to complete the disclosure of the invention. Since in almost all cases the deposit must have been effected at the latest by the filing date (or, where applicable, by the priority date), it might fairly be said that, apart from drafting the specification, this is often the first practical step to be taken towards obtaining the patent. Moreover, it is a step that relies for its effective accomplishment much more on the inventor than on his agent. The latter can do little more than advise about the documentary formalities and deadlines and perhaps suggest appropriate collections. It is the inventor who knows his organism, the technical difficulties in handling it, how long is needed to grow it, and any legal constraints in respect of its pathogenicity, which might delay matters. Thus it is up to the inventor to brief his agent so that between them they can ensure that the culture collection receives the organism in good time to allow for any possible delays or mishaps.

As mentioned above, the mechanism of deposit is now regulated internationally by the Budapest Treaty, and even in countries not yet party to the Treaty, its procedures tend to be viewed as a model system. Nevertheless, for purely national purposes, deposit under the Treaty is often not necessary (see National deposit requirements in Section 6.4.2 above). However, for the international recognition of a single deposit, using the Budapest Treaty is by far the safest course of action and the following account will be concerned mainly with the Budapest Treaty system. Although the following discussion goes into some detail, much more comprehensive information is contained in the *Guide to the Deposit of Microorganisms for the Purposes of Patent Procedure* issued by the World Intellectual Property Organization (WIPO), Geneva. For convenience, the term 'depositor' will be used in this connection in preference to 'applicant' or 'inventor'. Lastly, it should be borne in mind throughout that the date of deposit is the date on which the culture collection physically *receives* the culture, rather than the date when the culture is formally accepted.

Requirements of the Budapest Treaty. Under the Budapest Treaty a deposit must be made with an International Depositary Authority (IDA) according to the provisions of Rule 6 of the Treaty. The requirements for making such a deposit are laid down in Rule 6.1(a), which requires that the culture sent to an IDA must be accompanied by a written statement, signed by the depositor and containing the following information:

> (i)   an indication that the deposit is made under the Treaty and an undertaking not to withdraw it for the period specified in Rule 9.1;

The period specified in Rule 9.1 is five years after the latest request for a sample, and in any case at least 30 years. The important thing to note here is that a deposit made under the Budapest Treaty is permanent and having made it, the depositor cannot later ask for it to be cancelled, regardless of whether a patent is eventually granted. This applies even if he abandons his patent application.

> (ii)   the name and address of the depositor;
> (iii)   details of the conditions necessary for the cultivation of the microorganism, for its storage and for testing its viability and also, where a mixture of microorganisms is deposited, descriptions of the components of the mixture and at least one of the methods permitting the checking of their presence;

This requirement simply ensures that the culture collection is given enough information to enable it to handle the organism correctly. The instructions about (intentionally) mixed cultures are included so that a positive viability statement (see below) is not issued when all the components of the co-culture are not viable.

> (iv)   an identification reference (number, symbols, etc.) given by the depositor to the microorganism;

The term 'identification reference' is sometimes wrongly taken to refer to a taxonomic identification, whereas it simply means 'strain designation'.

> (v)   an indication of the properties of the microorganism which are or may be dangerous to health or the environment, or an indication that the depositor is not aware of such properties.

The requirements of Rule 6.1(a) are mandatory and cannot be varied either by the depositor or by the IDA. Indeed if the depositor does not comply with them all, the IDA is obliged to ask him to do so before it can accept the deposit. The same does not apply to Rule 6.1(b), which is not really a rule at all but simply an exhortation. According to Rule 6.1(b) 'it is strongly recommended that the written statement . . . should contain the scientific description and/or proposed taxonomic designation of the deposited microorganism'.

As well as the above requirements, the Treaty permits the IDA to set certain conditions of its own [Rule 6.3(a)]. These are:

> (i)   that the microorganism be deposited in the form and quantity necessary for the purpose of the Treaty and these Regulations;

Thus an IDA may require that cultures are submitted to it in a particular state, e.g. freeze-dried, in agar stabs, etc., and that a specified number of replicates is provided.

> (ii)   that a form established by such authority and duly completed by the depositor for the purposes of the administrative procedures of such authority be furnished;

This refers to the accession form (and any other form) routinely used by the culture collection.

> (iii)   that the written statement . . . be drafted in the language, or in any of the languages, specified by such authority . . .

This is an obvious requirement, permitting a Japanese depository, for example, to ask for information to be supplied to it in Japanese.

> (iv)   that the fee for storage . . . be paid;
> (v)   that, to the extent permitted by the applicable law, the depositor enter into a contract with such authority defining the liabilities of the depositor and the said authority.

This provides for the IDA to make the kind of contractual arrangements with the depositor that would be usual under the laws of contract of the IDA's own country. Without this provision, some culture collections would have been unwilling to become IDAs.

It is entirely up to the IDA whether it requires any or all of the above from the depositor, but if it does, then the depositor has no option but to comply. Some of the requirements of existing IDAs are summarised in Table 6.5.

These, then, are the official requirements which the depositor must meet. For its part, the IDA also must fulfil certain obligations under the Treaty. In particular it must issue to the depositor an official receipt (the contents of which are laid down by the Treaty) stating that it has received and accepted the deposit and it must, as soon as possible, test the viability of the culture deposited and issue an official statement to the depositor informing him of the result. If the culture proves not to be viable, the deposit is worthless, which can lead to major problems (see below). The IDA must also keep the deposit secret from all except those entitled to receive samples; it must maintain the deposit for the 30 or more years required by the Treaty, checking the viability 'at reasonable intervals' or at any time on the demand of the depositor; it must supply cultures to anyone entitled under the relevant patent law to receive

Table 6.5. *International Depositary Authorities at 1 January 1987*

International Depositary Authority	Microorganisms accepted	Minimum no. of replicates to be provided by the depositor
Summary		
Agricultural Research Collection (NRRL) Peoria USA	Non-pathogenic bacteria, actinomycetes, yeasts, moulds	
American Type Culture Collection (ATCC) Rockville USA	Most kinds	
Centraalbureau voor Schimmelcultures (CBS) Baarn Netherlands	Fungi, yeasts, actinomycetes, bacteria	
Collection Nationale de Cultures de Microorganismes (CNCM) Paris France	Bacteria, actinomycetes, fungi, yeasts, viruses	
Culture Collection of Algae and Protozoa (CCAP) Ambleside and Oban UK	Algae, non-pathogenic protozoa	
Culture Collection of the CAB International Mycological Institute (CMI CC) Kew UK	Non-pathogenic fungi	

Deutsche Sammlung von Mikroorganismen (DSM) Göttingen Federal Republic of Germany	Non-pathogenic bacteria, actinomycetes, fungi, yeasts, phages
European Collection of Animal Cell Cultures (ECACC) Salisbury UK	Cell lines, animal viruses
Fermentation Research Institute (FRI) Ibaraki-ken Japan	Non-pathogenic fungi, yeasts, bacteria, actinomycetes
In Vitro International Inc. (IVI) Linthicum USA	Most kinds
National Collection of Agricultural & Industrial Microorganisms (NCAIM) Budapest Hungary	Non-pathogenic bacteria, fungi, yeasts
National Collection of Yeast Cultures (NCYC) Norwich, UK	Non-pathogenic yeasts

Table 6.5. (cont.)

International Depositary Authority	Microorganisms accepted	Minimum no. of replicates to be provided by the depositor
Detailed information		
France		
Collection Nationale de Cultures de Micro-organismes (CNCM) Institut Pasteur 28 rue du Dr Roux 75724 Paris Cedex 15	Bacteria (including actinomycetes), bacteria containing plasmids; filamentous fungi and yeasts, and viruses, EXCEPT: – cellular cultures (animal cells, including hybridomes and plant cells); – microorganisms whose manipulation calls for physical insulation standards of P3 or P4 level, according to the information provided by the National Institutes of Health (NIH) Guidelines for Research Involving Recombinant DNA Molecules and Laboratory Safety Monograph; – microorganisms liable to require viability testing that the CNCM is technically not able to carry out; – mixtures of undefined and/or unidentifiable microorganisms. The CNCM reserves the possibility of refusing any microorganism for security reasons: specific risks to human beings, animals, plants and the environment.	Cell lines, 12 Other organisms, 8

France (*cont.*)

In the eventuality of the deposit of cultures that are not or cannot be lyophilized, the CNCM must be consulted, prior to the transmittal of the microorganism, regarding the possibilities and conditions for acceptance of the samples; however, it is advisable to make the prior consultation in all cases.

Federal Republic of Germany

Deutsche Sammlung von Mikroorganismen (DSM)
Gesellschaft für Biotechnologische Forschung mbH
Grisebachstr. 8
3400 Göttingen

Bacteria including actinomycetes, fungi, including yeasts, bacteriophages, except any kinds pathogenic to humans or animals. Phytopathogenic kinds are accepted, EXCEPT:

Erwinia amylovora; Coniothyrium fagacearum; Endothia parasitica; Gloeosporium ampelophagum; Septoria musiva; Synchytrium endobioticum.

2

Hungary

Mezőgazdasági és Ipari Mikroorganizmusok Magyar Nemzeti Gyüjteménye (MIMNG)
[National Collection of Agricultural and Industrial Microorganisms (NCAIM)]
Kertészeti Egyetem, Mikrobiológiai Tanszék
(Department of Microbiology, University of Horticulture)
Somlói ut 14–16
H-1118 Budapest

– Bacteria (including *Streptomyces*) except obligate human pathogenic species (e.g., *Corynebacterium diphtheriae, Mycobacterium leprae, Yersinia pestis,* etc.);
– Fungi, including yeasts and moulds, except some pathogens (*Blastomyces, Coccidioides, Histoplasma,* etc.) as well as certain basidiomycetous and plant pathogenic fungi which cannot be preserved reliably.

Apart from the above-mentioned, the following may not, at present, be accepted for deposit:

– viruses, phages, rickettsiae;
– algae, protozoa;
– cell lines, hybridomes.

3 or 25[a]

Table 6.5. (*cont.*)

International Depositary Authority	Microorganisms accepted	Minimum no. of replicates to be provided by the depositor
Japan		
Fermentation Research Institute (FRI) 1-3, Higashi 1-chome Yatabe-machi Tsukuba-gun, Ibaraki-ken 305	Fungi, yeast, bacteria and actinomycetes, EXCEPT: – microorganisms having properties which are or may be dangerous to health or the environment; – microorganisms which need the physical containment level P2, P3 or P4 required for experiments, as described in the 1979 Prime Minister's Guideline for Research Involving Recombinant DNA Molecules.	5
Netherlands		
Centraalbureau voor Schimmelcultures (CBS) Oosterstraat 1 Postbus 273 NL-3740 AG Baarn	Fungi, including yeasts; actinomycetes, bacteria other than actinomycetes.	6
United Kingdom		
Culture Collection of Algae and Protozoa (CCAP) Freshwater Biological Association Windermere Laboratory The Ferry House Far Sawrey Ambleside, Cumbria LA22 0LP and Scottish Marine Biological Association Dunstaffnage Marine Research Laboratory P.O. Box 3 Oban, Argyll PA34 4AD	(i) Freshwater and terrestrial algae and free-living protozoa (Freshwater Biological Association); and (ii) marine algae, other than large seaweeds (Scottish Marine Biological Association).	6

United Kingdom (*cont.*)

Culture Collection of the Mycological
   Institute
(CMI CC)
Ferry Lane
Kew
Surrey TW9 3AF

Fungal isolates, other than known human and animal pathogens and yeasts, that can be preserved without significant change to their properties by the methods of preservation in use.

European Collection of Animal Cell
   Cultures
(ECACC)
[formerly known as the National Collection of Animal Cell Cultures (NCACC)]
Vaccine Research and Production
   Laboratory
Public Health Laboratory Service
Centre for Applied Microbiology and
   Research
Porton Down
Salisbury, Wiltshire SP4 0JG

Cell lines that can be preserved without significant change to or loss of their properties by freezing and long term storage; viruses capable of assay in tissue culture. A statement on their possible pathogenicity to man and/or animals is required at the time of deposit. Up to and including ACDP Category 3 can be accepted for deposit (Advisory Committee on Dangerous Pathogens: Categorisation of Pathogens according to Hazard and Categories of Containment ISBN 0/11/883761/3 HMSO London).

12, each containing at least 2×10^6 cells

National Collection of Industrial Bacteria
(NCIB)
c/o The National Collections of Industrial
and Marine Bacteria Ltd.
Torry Research Station
P.O. Box 31
135 Abbey Road
Aberdeen AB9 8DG

(a) Bacteria, including actinomycetes, that can be preserved without significant change to their properties by liquid nitrogen freezing or by freeze-drying (lyophilisation), and which are allocated to a hazard group no higher than Group 2 as defined by the UK Advisory Committee on Dangerous Pathogens (ACDP);

(b) Plasmids, including recombinants, either

   (i) cloned into a bacterial or actinomycete host, or

   (ii) as naked DNA preparations.

As regards (i) above, the hazard category of the host with or without its plasmid must be no higher than ACDP Group 2.

Bacteria and phages, 2
Naked plasmids, 20

Table 6.5. (cont.)

International Depositary Authority	Microorganisms accepted	Minimum no. of replicates to be provided by the depositor
United Kingdom (cont.)	As regards (ii), above, the phenotypic markers of the plasmid must be capable of expression in a bacterial or actinomycete host and must be readily detectable. In all cases, the physical containment requirements must not be higher than level II as defined by the UK Genetic Manipulation Advisory Group (GMAG) and the properties of the deposited material must not be changed significantly by liquid nitrogen freezing or freeze-drying. (c) Bacteriophages that have a hazard rating and containment requirement no greater than those cited in (a) or (b), above, and which can be preserved without significant change to their properties by liquid nitrogen freezing or by freeze-drying. Notwithstanding the foregoing, the NCIB reserves the right to refuse to accept any material for deposit which in the opinion of the Curator presents an unacceptable hazard or is technically too difficult to handle. In exceptional circumstances the NCIB may accept deposits which can only be maintained in active culture, but acceptance of such deposits, and relevant fees, must be decided on an individual basis by prior negotiation with the prospective depositor.	

United Kingdom (cont.)

National Collection of Type Cultures
(NCTC)
Central Public Health Laboratory
175 Colindale Avenue
London NW9 5HT

Bacteria that can be preserved without significant change to their properties by freeze-drying and which are pathogenic to man and/or animals.

1

National Collection of Yeast Cultures
(NCYC)
Food Research Institute
Colney Lane
Norwich, Norfolk NR4 7UA

Yeasts other than known pathogens that can be preserved without significant change to their properties by freeze-drying or, exceptionally, in active culture.

1

United States of America

American Type Culture Collection
(ATCC)
12301 Parklawn Drive
Rockville, Maryland 20852

Algae, animal viruses, bacteria, cell lines, fungi, hybridomas, oncogenes, phages, plant tissue cultures, plant viruses, plasmids, protozoa, seeds, yeasts.

The ATCC must be informed of the physical containment level required for experiments using the host vector system, as described in the 1980 National Institutes of Health Guidelines for Research involving Recombinant DNA Molecules (i.e. P1, P2, P3 or P4 facility). The ATCC, for the time being, will accept only those hosts containing plasmids which can be worked in a P1 or P2 facility.

Certain animal viruses may require viability testing in an animal host, which the ATCC may be unable to provide. In such cases, the deposit cannot be accepted. Plant viruses which cannot be mechanically inoculated also cannot be accepted.

Animal viruses
Cell lines } 25
Naked plasmids
Other organisms, } 6
Seeds, 250

Table 6.5. (cont.)

International Depository Authority	Microorganisms accepted	Minimum no. of replicates to be provided by the depositor
United States of America (cont.) Agricultural Research Service Culture Collection (NRRL) 1815 North University Street Peoria, Illinois 61604	Progeny of strains of agriculturally and industrially important bacteria, yeast, molds, and *Actinomycetles*, EXCEPT: (a) *Actinobacillus* (all species); *Acytomyces* (anaerobic/microaerophilic – all species); *Arizona* (all species); *Bacillus anthracis*; *Bartonella* (all species); *Bordetella* (all species); *Borrelia* (all species); *Brucella* (all species); *Clostridium botulinum*; *Clostridium chauvoei*; *Clostridium haemolyticum*; *Clostridium histolyticum*; *Clostridium novyi*; *Clostridium septicum*; *Clostridium tetani*; *Corynebacterium diphtheriae*; *Corynebacterium equi*; *Corynebacterium haemolyticum*; *Corynebacterium pseudotuberculosis*; *Corynebacterium pyogenes*; *Corynebacterium renale*; *Diplococcus* (all species); *Erysipelothrix* (all species); *Escherichia coli* (all enteropathogenic types); *Francisella* (all species); *Haemophilus* (all species); *Herellea* (all species); *Klebsiella* (all species); *Leptospira* (all species); *Listeria* (all species); *Mima* (all species); *Moraxella* (all species); *Mycobacterium avium*; *Mycobacterium bovis*; *Mycobacterium tuberculosis*; *Mycoplasma* (all species); *Neisseria* (all species); *Pasteurella* (all species); *Pseudomonas pseudomallei*; *Salmonella* (all species); *Shigella* (all species); *Sphaerophorus* (all species); *Staphylococcus*	1 or 30[a]

United States of America (*cont.*)

aureus; Streptobacillus (all species); *Streptococcus* (all pathogenic species); *Treponema* (all species); *Vibrio* (all species); *Yersinia* (all species);

(b) *Blastomyces* (all species); *Coccidioides* (all species); *Cryptococcus* (all species); *Histoplasma* (all species); *Paracoccidioides* (all species);

(c) *Basidiomycetes* or other molds that cannot successfully be preserved by lyophilization (freeze-drying);

(d) all viral, Rickettsial, and Chlamydial agents;

(e) agents which may introduce or disseminate any contagious or infectious disease of animals, humans, or poultry and which would require a permit for entry and/or distribution within the United States of America;

(f) agents which are classified as Plant Pests and which would require a permit for entry and/or distribution within the United States of America;

(g) mixtures of microorganisms;

(h) fastidious microorganisms which would require (in the view of the Curator) more than reasonable attention in handling and preparation of lyophilized material;

(i) phage of any kind;

(j) plasmids and like materials.

Table 6.5. (cont.)

International Depositary Authority	Microorganisms accepted	Minimum no. of replicates to be provided by the depositor
United States of America (cont.) In Vitro International, Inc. (IVI) 611(P) Hammonds Ferry Road Linthicum, Maryland 21090	Algae, bacteria with plasmids, bacteriophages, cell cultures, fungi, protozoa and animal and plant viruses. Recombinant strains of microorganisms will also be accepted, but IVI must be notified in advance of accepting the deposit of the physical containment level required for the host vector system, as prescribed by the National Institutes of Health Guidelines. At present, IVI will accept only hosts containing recombinant plasmids that can be worked in a P1 or P2 facility.	Bacteria ⎤ Fungi ⎬ 3 Yeasts ⎦ Seeds, 400 Other organisms, 6

[a] If depositor's own lyophilised cultures are to be stored and distributed.

them (provided that the IDA has been given proof of entitlement – see below); it must inform the depositor when and to whom it has released samples; it must be impartial and available to any depositor under the same conditions.

New deposits. If by some mischance a microorganism which was viable when deposited dies during storage, or if indeed for any reason the IDA can no longer supply cultures of it, then the IDA must notify the depositor immediately. The latter then has the option of replacing it (Article 4), and provided he does so within three months, the date on which the original deposit was made still stands. When making a new deposit, the depositor must (under Rule 6.2) provide the IDA with:

(1) a signed statement that he is submitting a culture of the same microorganism as deposited previously;
(2) an indication of the date on which he received notification from the IDA of its inability to supply cultures of the previous deposit;
(3) the reason he is making the new deposit;
(4) a copy of the receipt and the last positive viability statement in respect of the previous deposit;
(5) a copy of the most recent scientific description and/or taxonomic designation submitted to the IDA in respect of the previous deposit;
(6) if the new deposit is being made with a different IDA, all the indications required under Rule 6.1(a) (see above).

With regard to item (6) above, the new deposit can be made with a different IDA if the original IDA is no longer operating as such (either entirely or just in respect of that particular kind of microorganism) or if import/export regulations render the original IDA inappropriate for that particular deposit.

It must be remembered that the provisions for making a new deposit cannot be applied to a microorganism which was shown by the IDA to be non-viable when it was originally deposited. There must have been at least one positive viability statement.

'Converted' deposits. The Budapest Treaty allows [Rule 6.4(d)] for a deposit made outside its provisions to be 'converted' to a Treaty deposit (provided, of course, that the culture collection holding the deposit is an IDA). Where the microorganism was deposited before the culture collection became an IDA the date of deposit of the 'conversion' is held

to be the date on which the collection acquired IDA status. Otherwise the date of deposit is the date on which the collection physically received the culture. The procedure for converting a deposit usually involves completing the same forms as are used for making a deposit *de novo*. However, only the original depositor (or his successor) can convert a deposit. In all other cases, a separate deposit of the same organism must be made under the Treaty.

Conversion is a useful facility because it means that an earlier non-Budapest Treaty deposit can be accorded the international recognition which it might not otherwise command. Conversion is essential for the recognition by the Japanese patent office of any non-Budapest Treaty deposit made outside Japan.

Guidelines for deposit. It is generally recognised that many depositors (and sometimes their patent agents) are unlikely to be familiar with the minute details of the Budapest Treaty and may not be aware of their obligations in respect of it. Therefore, the forms which IDAs ask prospective depositors to fill in are generally so designed that by completing them correctly the depositor automatically provides all the information required of him by the Treaty. These forms vary to some extent between IDAs, but they all follow a similar general pattern. Any IDA will supply specimens of its forms on request.

Making a deposit under the Budapest Treaty should be quite straight-forward, but problems can and do arise. It has to be said that many of these are of the depositor's (or his agent's) own making and that they can be avoided by adhering to a few simple guidelines. Perhaps the most important thing to be remembered is that the Budapest Treaty procedures take a certain amount of time to complete, even when they are operating ideally. Thus although in principle a deposit does not have to reach the IDA until the filing (or priority) date of the relevant patent application, in practice the wise depositor will start the depositing procedure in good time to allow for any possible delays. Thus if he is intending to deposit in a foreign IDA, say, he should bear in mind any import or quarantine regulations. For instance, it can take several weeks, or even months, to obtain a permit to import cell lines and viruses into the USA. Last-minute deposits are unwise for several reasons, some of the most common being:

(1) postal delays: the culture fails to arrive in time;
(2) customs delays: with deposits from overseas, depositors have not provided adequate shipping information;

(3) the deposit is not the kind of microorganism accepted by the IDA (see Table 6.5);

(4) the microorganism cannot be recovered from the package, e.g. because the culture tube is broken;

(5) the deposit proves to be non-viable; if a microorganism is found by the IDA from the outset to be non-viable, the original date of deposit cannot be applied to any replacement (see above).

For most bacteria, fungi, yeasts, algae and protozoa, viability testing usually takes 3–5 days; for animal cell lines a week or slightly longer is normal; and for animal viruses and plant tissue cells, up to a month is not unusual.

It cannot be emphasised too strongly that however good the depositor's intentions may be, patent offices recognise only the actuality of the deposit. With all this in mind the prudent depositor will also pay attention to a few more elementary points to ensure a timely and trouble-free deposit. He will ensure that the microorganism he wishes to deposit is one of the kinds that the IDA he had chosen can officially accept under the Budapest Treaty (see Table 6.5). If there are likely to be technical problems with the organism he will advise the IDA in advance. He will check the administrative and technical requirements of the chosen IDA and ask for the appropriate patent deposit forms, which he will then fill in completely and correctly, for by doing so he should automatically comply with the requirements of Rule 6.1(a) (see above). Although Rule 6.1(a) states that the microorganism should be *accompanied* by a written statement (the completed deposit form), in practice it is often helpful to an IDA to receive the written information in advance of the microorganism itself, so that arrangements can be made to deal with the deposit promptly. This is particularly helpful if, say, a special growth medium has to be prepared by the IDA. Lastly, if the depositor's patent agent is likely to be communicating with the IDA, the depositor should let the IDA know, otherwise it may withhold information until it has ascertained the agent's right to receive it.

Depending on its policy and on the kind of material being deposited, an IDA may or may not prepare subcultures for eventual distribution. Thus in the case of cell lines and naked plasmids (not cloned into a host), for instance, the depositor is usually required to supply sufficient material for the IDA to distribute direct. On the other hand, for bacteria, yeasts, moulds, etc. (with or without plasmids), it is more usual for the IDA to distribute its own preparations. In this case, many IDAs will ask the depositor to check the authenticity of their preparations – a fairly normal culture collection practice. The depositor is not *obliged* by the Budapest Treaty to check these preparations, but he is well advised to do so to ensure that the cultures to be sent out by the IDA will in fact do what is claimed for them in the patent application.

The official aspects of the depositing procedure end with the issuing of the receipt and viability statement by the IDA. These are important for they are the documentary proof that on a particular date a viable deposit has been made according to the terms of the Budapest Treaty.

Technically the receipt should be issued first, but in practice many IDAs find it more convenient to await the results of the viability test and then send out the receipt and viability statement together. In general, for deposits of most bacteria, fungi, yeasts, algae and protozoa, the depositor could expect an IDA to send him both documents within a few days of it having received the deposit. For animal cell lines a week or slightly longer would be normal, and for animal viruses and plant tissue cells four or five weeks would be more usual.

6.5.6   *Obtaining a sample of a patent deposit*

So far in this account the deposit procedure has been considered primarily from the viewpoint of the depositor. It would be useful now to look briefly at the procedures which a third party must follow in order to obtain a culture, since the whole point of deposit is to make the microorganism available.

It is generally admitted that culture collections can neither be expected to be familiar with the patent laws of countries throughout the world nor to know what stage patent applications relating to the deposits they hold have reached. Thus to require a collection to judge for itself whether a particular person is legally entitled to a culture of a particular deposit is considered by many to impose an unfair burden on the collection. Therefore the Budapest Treaty attempts to place the onus on patent offices to ensure that IDAs are not put in this position (Rule 11.3). Patent offices in countries whose laws require that deposited microorganisms must be available without restriction to anyone once the relevant patents have been granted and published can notify the IDAs from time to time of the accession numbers of the strains cited in these patents. However, this provision is not usually adopted. The US Patent Office, for instance, directs that a microorganism must be available from the date of issuance of the relevant US patent, but it does not advise the IDA of this date. Since the issuance or not of a US patent is a simple matter of fact, in the event of any request an IDA merely has to ascertain this fact, either from the requesting party or the depositor. In the author's experience this does not cause any major problems, although the actual furnishing of the sample may be delayed slightly while evidence of publication is obtained. The IDA can then meet any requests for the strains in question without the need for evidence of entitlement.

In cases where the availability of the microorganism is restricted and/or where evidence of entitlement to receive a sample is required, anyone requiring a culture must either obtain the written authorisation of the depositor or he must obtain from a relevant patent office a certificate stating:

(1) that a patent application in respect of the strain in question has actually been filed with that office;
(2) whether the application has been published;
(3) that the person requesting the culture is legally entitled to receive it and has met any conditions that the law requires.

On receipt of a request accompanied by such a certificate, or by the written authorisation of the depositor, the IDA will supply the culture (subject to its normal fee for such cultures being paid). At the same time, the IDA will inform the depositor when and to whom it has supplied the culture, as it is obliged to do by Rule 11.4(g) of the Treaty, unless the depositor has specifically waived his right to be informed.

Except where the direct authorisation of the depositor has been sought, the request for a culture must be made on an official form which can be had from the patent office(s) with which the relevant application has been filed. Most IDAs also have copies of these forms. Thus, the procedure for obtaining a culture of a microorganism deposited under the Budapest Treaty is:

(1) ask the appropriate patent office, or the IDA, for a copy of the form to be used for requesting samples of microorganisms deposited under the Budapest Treaty;
(2) complete that part of the form to be filled in by 'the requesting party';
(3) send the entire form to the patent office, *not* to the IDA;
(4) when the form bearing the appropriate stamp of authorisation is received back from the patent office, send it to the IDA along with a normal purchase order.

Procedures for obtaining organisms deposited for patent purposes outside the Budapest Treaty vary according to the national law. In such cases, the culture collection will have been informed by the depositor or the patent office of the appropriate requirements and should be able to advise accordingly.

It must be remembered that the procedures outlined above relate only to the right to receive cultures according to patent law. They do not override any requirements to be met in respect of import and quarantine regulations, health and safety procedures, plant disease regulations, etc. Thus as well as obtaining patent office authorisation, a person requesting a culture must also ensure that he has obtained any permit or licence necessary for handling the organism in question.

6.6   Further reading

Adler, R. G. (1984). Biotechnology as an intellectual property. *Science* **224**, 357–63.

Anonymous (1982). Japanese Patent Office guidelines for examination of inventions of microorganisms. *Yuasa and Hara Journal* **9** (3).

Biggart, W. A. (1981). Patentability in the United States of microorganisms, processes utilizing microorganisms, products produced by microorganisms and microorganism mutational and genetic information techniques. *IDEA, Journal of Law and Technology* **22**, 113–36.

Byrne, N. J. (1979). Patents on life. *European Intellectual Property Review* **1**, 279–300.

Byrne, N. J. (1983). The agritechnical criteria in plant breeders' rights law. *Industrial Property* (1983), 294–303.

Cooper, I. P. (1902). *Biotechnology and the Law*. New York: Clark Boarman Co.

Crespi, R. S. (1981). Biotechnology and patents – past and future. *European Intellectual Property Review* **3**, 134–40.

Crespi, R. S. (1985). Biotechnology patents – a case of special pleading? *European Intellectual Property Review* **7**, 190–3.

Crespi, R. S. (1985). Microbiological inventions and the patent law – the international dimension. *Biotechnology and Genetic Engineering Reviews* **3**, 1–37.

Crespi, R. S. (1986). Patent issues in biotechnology. In *Biotechnology and Crop Improvement and Protection*, British Crop Protection Council Monograph No. 34, ed. Peter R. Day, pp. 209–17.

Halluin, A. P. (1982). Patenting the results of genetic engineering research: an overview. In *Patenting of Life Forms*, Banbury Report No. 10, pp. 67–126. Cold Spring Harbor, New York: Cold Spring Harbor Laboratory.

Hüni, A. (1977). The disclosure in patent applications for microbiological inventions. *International Review of Industrial Property and Copyright Law* **8**, 500–21.

Hüni, A. & Buss, V. (1982). Patent protection in the field of genetic engineering. *Industrial Property* (1982), 356–68.

Irons, E. S. & Sears, M. H. (1975). Patents in relation to microbiology. *Annual Review of Microbiology* **29**, 319–32.

Plant, D. W., Reimers, N. J. & Zinder, N. D. (eds.) (1982). *Patenting of Life Forms.* Banbury Report No. 10. Cold Spring Harbor, New York: Cold Spring Harbor Laboratory.

Pridham, T. G. & Hesseltine, C. W. (1975). Culture collections and patent depositions. *Advances in Applied Microbiology* **19**, 1–23.

Teschemacher, R. (1982). Patentability of microorganisms *per se. International Review of Industrial Property and Copyright Law* **13**, 27–41.

Wegner, H. C. (1979). Patenting the products of genetic engineering. *Biotechnology Letters* **1**, 145–50 and 193.

Wegner, H. C. (1980). The Chakrabarty decision patenting products of genetic engineering. *European Intellectual Property Review* **2**, 304–7.

The author gratefully acknowledges the helpful comments and criticisms made by many colleagues during the writing of this chapter. Particular thanks go to Mrs B. A. Brandon of the American Type Culture Collection and Mr R. S. Crespi of the British Technology Group. A singular debt of gratitude is owed to Mr R. K. Percy of the British Technology Group, for his extensive advice and painstaking correction of the original draft.

7

Culture collection services

D. ALLSOPP and F. P. SIMIONE

7.1   Introduction

In response to the needs of users, many culture collections provide a range of services to the scientific, technological and commercial world. This chapter provides an introduction to the types of services available from culture collections, but it is beyond its scope to give a comprehensive list of such services. As the range of work that can be undertaken is increasing at many of the collections, the reader should contact individual collections to find out whether they can offer particular services.

7.2   Types of services

7.2.1   *Directly associated and customer services*

The two major services which are intrinsically part of culture collection work are those concerning the identification and preservation of organisms. Collections of necessity need expertise in these fields to be able to function, and many provide comprehensive services in these areas. Aspects of culture identification methods (Chapter 5), sales of cultures (Chapter 3), preservation techniques (Chapter 4) and patent deposits (Chapter 6) are covered elsewhere in this volume.

Safe-deposits. Many collections hold organisms which are not listed in their catalogues. These cultures are held for a variety of reasons: the organisms may not be fully identified, their taxonomic status may be unclear, their stability in preservation may be suspect or they may be held at the request of the depositor who wishes to have back-up material and yet retain ownership and confidentiality, not releasing the strain to other parties. Many collections have introduced safe-deposit services as a back-up to the depositor's working collection, providing a service intermediate between an open collection deposit and a deposit for patent purposes. Such services enable the depositor to have important organisms professionally preserved and maintained even if the collection would not normally be interested in accessioning them. There is obvious merit in the safekeeping of cultures while they are the subject of research, especially as many laboratories do not have optimum preservation facilities. Most collections make a charge for safe-deposits to cover the long-term storage costs and quality control procedures that are required.

Advice on strain selection. Collections are able to give advice on the selection of strains for special purposes. Such services may involve collection staff in a considerable amount of work and are limited by the sources of information available. In the past such services have relied upon the individual expertise of collection staff and their personal knowledge of the scientific literature. With the development of computer databases and strain data networks (Chapter 2), such services are becoming more frequent and are increasing in efficiency. However, as databases are searched electronically and appropriate microorganisms selected, the expertise of the collection staff is still needed to draw attention to closely allied genera and species that might merit study or to point out the idiosyncracies of individual strains. These advisory services are being placed on a more formal basis and charges may be made.

Advice on maintenance of organisms. Most collections are able to provide information on preservation systems, either on a formal or informal basis. Some of the major collections have produced substantial publications covering the preservation of their own groups of organisms and these may be consulted. However, if any doubt exists or if different media are being tried, the collections can always be consulted for advice. Advisory sheets on particular preservation techniques, or related topics such as the handling of pathogens, elimination of contamination or mite infestation, are often available on request.

Biochemical services. Some groups of organisms, such as bacteria and yeasts, are identified using biochemical tests, and such methods are increasingly used for other organisms, particularly the filamentous fungi. The need for taxonomic clarification using biochemical criteria goes hand in hand with an increasing requirement to provide metabolic and other physiological data to users of the collections. Applied biological industries, and in particular biotechnology, increasingly seek organisms on the basis of activity rather than name, and collections are responding to this requirement.

Collections now enhance their catalogues with strain data, and supply information to strain databases and computer networks as a routine procedure. Some collections also carry out custom-designed screening programmes to select strains with specified attributes for individual clients, particularly in the fields of enzyme and secondary metabolite (including toxin) production or specific growth requirements.

These biochemical services serve both as directly associated and contract services of culture collections (see also Research and development work in section below).

7.2.2   Contract services

The services outlined in Section 7.2.1 above usually exist as a consequence of the collection's normal activities; however, in recent years there has been an expansion in other services offered by collections for a variety of reasons. For example, many collections are associated with other institutions, such as research organisations, taxonomic institutes, educational institutions, or commercial firms, any of which may make specific demands on the services of the collection. Again, the collection may be supported by funds provided from external sources which may require particular expertise or services to be developed. Many collections are currently under financial pressure to increase their earning potential, and this has stimulated the development of income-generating activities.

Biological testing. Many modern standards and specifications require the use of microorganisms and cell lines in the testing of products. These include the mould growth resistance of materials (Fig. 7.1), the testing of disinfectants against a range of bacteria, the assessment of mutagenicity of materials against a range of organisms, and toxicity testing. For economic and other reasons there is strong pressure to move away from the use of live animals in the testing of products and towards the use of microorganisms or cell lines instead.

These tests can be carried out in any suitably equipped laboratory, but culture collections are in a very favourable position to carry out such testing themselves, using the organisms or cell lines which they would

Fig. 7.1. Chamber used for the testing of industrial products for resistance to mould growth at the CAB International Mycological Institute (IMI).

normally supply to outsiders for such work; indeed, collections may be identified in the standards themselves as an approved source of such material. Unless a company is equipped for routine testing as part of its quality control procedures, it is often more cost effective for such work to be carried out in a specialised laboratory. To establish a laboratory and train staff to carry out extensive biological testing infrequently would be very expensive, and not necessarily satisfactory from a technical point of view. However, an economic service can be offered by culture collections equipped to carry out such work on a regular basis.

Examples of testing standards using microorganisms and cell lines include British (BS), European (ISO) and USA (ASTM, USP) standards for fungal resistance testing, sterility testing, preservative effectiveness, toxicity and biocompatibility testing (Table 7.1). Culture collections provide reference cultures used in clinical laboratory standards procedures (e.g. NCCLS, ECCLS), and can provide uniform, quality assured sets of reference cultures for use in on-site biological testing for both clinical and industrial applications.

Consultancy. Many collections have staff with expertise in specialised areas commensurate with their collection responsibilities who can be made available for consultancy work. Such work is often a forerunner of detailed investigations, or a research programme, and can be time consuming. Since collection staff time for this work is limited, consultancy work is usually offered on a fully charged basis.

Research and development work. Culture collections equipped for testing work and consultancy may also be involved in research and development work and in industrial investigations which require laboratory facilities. It is often difficult for collections to specify exactly what kind of work they would be prepared to accept, as many problems are unique and may be carried out on a one-off basis. It is usual, therefore, for collections to consider any type of work which falls in their general area of competence. By their very nature, culture collections usually have a wide range of scientific and industrial contacts, and even if they are not able to carry out particular investigations themselves, they may be able to advise on places where such work could be carried out. Culture collections can therefore act as referral centres and this function should not be overlooked.

In addition to research topics on taxonomy and the preservation of organisms, major culture collections are often well placed to undertake

Table 7.1. *Examples of testing standards involving microorganisms, cell cultures and related materials*

British Standards	
BS 1982	Methods of testing for fungal resistance of manufactured building materials made of, or containing, materials of organic origin
BS 2011	The environmental testing of electronic components and electronic equipment. Test J. Mould growth
BS 28458	Flexible insulating sleeving for electrical purposes. Section 12. Mould growth
BS 3046	Specification for adhesives for hanging flexible wall coverings. Appendix G. Test for susceptibility to mould growth
BS 4249	Specification for paper jointing. Section 5.7. Resistance to mould growth
BS 5980	Specification for adhesives for use with ceramic tiles and mosaics. Section 7. Resistance to mould growth
BS 6009	Wood preservatives. Determination of toxic values against wood destroying *Basidiomycetes* on an agar medium
BS 6085	Methods of test for the determination of the resistance of textiles to microbiological deterioration
European Standards	
ISO 846	Plastics – determination of behaviour under the action of fungi and bacteria – evaluation or measurement of change in mass or physical properties

See also NFX41–514 and DIN 53739 for similar methods of plastics testing in France and Germany, respectively.

US Standards	
ASTM D2574	Standard test method for resistance of emulsion paints in the container to attack by microorganisms
ASTM D3273	Resistance to growth of mould on the surface of interior coatings in an environmental chamber. Standard test method for evaluating the degree of surface disfigurement of paint films by fungal growth or soil and dirt contamination
ASTM G21–70	Standard recommended practice for determining resistance of synthetic polymeric materials to fungi
ASTM G22–76	Standard recommended practice for determining resistance of plastics to bacteria
FDA 21CFR610.12	Sterility testing of biological products
FDA 21CFR610.30	Detection of mycoplasma contamination
MIL STD 810D	Environmental test methods. Method 508.2 Fungus
NCCLS M2–A3	Performance standards for antimicrobial disk susceptibility tests

Table 7.1. (*cont.*)

NCCLS M7A	Methods for dilution antimicrobial susceptibility tests for bacteria that grow aerobically
NCCLS M11A	Reference agar dilution procedure for antimicrobial susceptibility testing of anaerobic bacteria
USDA 9CFR113.26	Detection of viable bacteria and fungi in biological products
USDA 9CFR113.28	Detection of Mycoplasma contamination
USDA 9CFR113.29	Determination of moisture content in desiccated biological products
USP	Antimicrobial preservatives – effectiveness assay
USP	Microbial limits tests
USP	Sterility tests

other research, often allied to biotechnology, which may be initiated by short-term contracts and testing work. Potential topics for such longer-term research and development work might include:

(1) screening large numbers of isolates for particular biochemical properties and end uses;

(2) comparative studies in regard to enzyme or metabolite production of different strains of the same or closely related species;

(3) development and evaluation of rapid detection procedures for compounds produced by organisms, including the development of commercial kits;

(4) studies on the use of microorganisms as bio-control agents against insect pests, weeds and deteriogenic microorganisms;

(5) selection of test organisms used in the evaluation of materials;

(6) studies on growth requirements and the testing of bioreactors;

(7) evaluation of media, culture vessels, diagnostic reagents and procedures.

Custom preparations. Culture collections may be requested to provide bulk amounts of their organisms on an *ad hoc* basis to commercial organisations with limited facilities. Often this is an efficient and cost-effective method of obtaining bulk inocula, since the collection has the expertise in growing the organisms and ready access to the media.

Government agencies may request multiple units of mixtures of cultures for laboratory certification and proficiency testing in clinical laboratories. Manufacturers of diagnostic instruments may specify cultures with known properties for calibration of their instruments.

Resource development. In the United States the Federal Government has used culture collections extensively to develop research reagents/ cultures. These have been developed through joint efforts with the research scientists defining the needs and quality controls required and the banking and distribution aspects handled by the collections.

7.2.3 *Confidentiality*

Work performed by culture collections can be carried out on a confidential basis if required. Formal mechanisms for ensuring confidentiality are in fact required by some laboratory accreditation schemes (see below) and are already in place in a number of collections. In some areas, such as identification services, users are sometimes content for material sent in to become part of the open collection, should the collection wish to retain it. However, in collections which deal with organisms of industrial importance, it may be more normal to carry out the work in confidence, with all material being destroyed after examination. Enquiries can be treated as confidential, and collections should be questioned about their policy and procedures regarding confidentiality. Submissions are a source of new organisms for building up the resources available from the collections, and some collections will identify cultures for a lower fee if the culture can be retained in the collection.

7.2.4 *Laboratory accreditation*

In many countries national schemes exist for the accreditation of testing laboratories. Such schemes aim to establish standards for the accuracy and efficiency of measuring instruments, quality control procedures, record-keeping, and administration of the laboratory. Major companies may have their own accreditation schemes for laboratories that they use, but where national schemes exist, companies often accept the accreditation of the scheme and do not carry out individual laboratory accreditation on their own behalf. National schemes may also be recognised on an international basis and thus ease the way for laboratories to be accepted on a much wider geographical basis. There are now a sufficient number of national and international testing standards involving microorganisms to make such accreditation schemes worthwhile for culture collection laboratories to adopt.

7.3 **Workshops and training**

Many culture collection staff are involved in educational activities, either because of their general scientific background or their detailed knowledge of culture collection practices. Instruction may be external, where staff are involved in training courses at local colleges, polytechnics or research institutes in their own country or overseas. Increasingly, however, collections are devising formal programmes of in-house courses. These may be arranged in direct response to a need identified either by the collection itself or by an outside body. In some parts of the world the educational system is such that credits towards a qualification can be gained by attendance at approved outside courses, and where such a system exists, it can ease the task of the culture collection in operating such courses.

Courses vary in length from several weeks' or even months' duration (if dealing with topics such as identification), to one-day lecture courses or seminars on specific industrial or commercial topics. Outside speakers and instructors may be used to enable topics to be covered which extend beyond the expertise available in the collection itself. In addition to formal programmed and advertised courses, it is often possible to obtain individual training within culture collections to suit a particular need. Some collections have special facilities and staff devoted to training programmes, while others may use external facilities at educational institutions nearby. Charges are usually made for training, but in some cases the costs may be subsidised. Advice may be available from collections on suitable sources of funding for prospective students and trainees, especially from developing countries. Some culture collections are either owned by academic institutions such as universities, or are officially associated with them, enabling them to offer training towards MSc and PhD degrees by research.

Some of the training offered is directly related to the normal activities of the collection, and instruction in preservation and maintenance techniques is often available; indeed, such training is often difficult to obtain from other sources. The training facilities offered by culture collections are often much greater than is generally known and the World Federation for Culture Collection's Education Committee is developing a list of teachers, their special expertise and courses available throughout the world (see Chapter 8). Individual collections may also provide information on training facilities in their scientific speciality or geographical area.

7.4   Publications, catalogues and publicity material

The essential publication of any culture collection is its catalogue. Traditionally, these have been produced as hard-copy items but are now becoming available in a computerised on-line form (see Chapter 2). It is always worthwhile contacting a collection if an organism does not appear in its most recent hard-copy catalogue, as additional material may well be available, or information can be given on suitable organisms in reserve collections, which could possibly be released. In addition to catalogues, a few of the major collections produce scientific publications of their own (guides on preservation and maintenance, safety and handling, industrial uses and teaching) as well as articles in the scientific press and monographic studies. Details of such publications may be found in the collection's brochures or newsletters, or in bibliographic databases.

Collection brochures are generally available free of charge, and it is worthwhile asking to be included on the collection's mailing list to ensure receipt of up-to-date information. Despite the fact that collections grow and change with time, enquirers or customers often rely on data from back issues of catalogues which may be many years old. This can lead to considerable confusion and users should ensure that they have the current catalogues before ordering cultures. These problems are minimised as catalogue and strain data become widely available through computer networks. For many people, however, printed information remains the most important reference material.

7.5   Fees and charges

The ways in which culture collections are funded are extremely diverse. Very few culture collections exist as straightforward commercial entities; they are almost all subsidised in some fashion, either directly or indirectly, and the charges made for cultures do not reflect the true cost of production. Nevertheless, many collections offer discounts for bulk orders, special sets, or regularly ordered organisms such as those used for testing and teaching.

The charges made for services, however, more accurately reflect the true cost, though they usually represent good value in comparison with totally commercial services, particularly in the areas of training. Charges for consultancy work, for testing and for laboratory services are normally at competitive commercial rates.

Over the last few years, several important stimuli have been applied to culture collections, including the advent of biotechnology, the development of computerised databases and a harsher economic climate. These and other factors have led culture collections to examine the services they provide and to develop them to cater to the changing demand of their users. Expansion and diversification of the range of services has resulted.

7.6   Suggested reading

Alexander, M., Daggett, P.-M., Gherna, R., Jong, S., Simione, F. & Hatt, H. (1980). *American Type Culture Collection Methods. Laboratory Manual on Preservation: Freezing and Freeze Drying*. Rockville, Maryland: American Type Culture Collection.

Allsopp, D. (1985). Fungal culture collections for the biotechnology industry. *Industrial Biotechnology* **5**, 2.

Allsopp, D. & Seal, K. J. (1986). *Introduction to Biodeterioration*, 136 pp. London: Edward Arnold.

Batra, L. R. & Iijima, T. (eds) (1984). *Critical Problems for Culture Collections*, 71 pp. Osaka, Japan: Institute for Fermentation.

Cour, I. G., Maxwell, G. & Hay, R. (1979). Tests for bacterial and fungal contaminants in cell cultures as applied at the ATCC. *TCA Manual* **5**, 1157–60.

Dilworth, S., Hay, R. & Daggett, P.-M. (1979). Procedures in use at the ATCC for detection of protozoan contaminants in cultured cells. *TCA Manual* **5**, 1107–10.

Hawksworth, D. L. (1985). Fungus culture collections as a biotechnological resource. *Biotechnology and Genetic Engineering Reviews*, **3**, 417–53.

Hay, R. J. (1983). Availability and standardization of cell lines at the American Type Culture Collection: Current status and prospects for the future. In *Cell Culture Test Methods*, STP 810, ed. S. A. Brown, pp. 114–26. Philadelphia: American Society for Testing and Materials.

Jewell, J. E., Workman, R. & Zelenick, L. D. (1976). Moisture analysis of lyophilized allergenic extracts. In *International Symposium on Freeze-Drying of Biological Products. Developments in Biological Standardization* **36**, 181–9.

Kelley, J. (1985). The testing of plastics for resistance to microorganisms. In *Biodeterioration and Biodegradation of Plastics and Polymers*, ed. K. J. Seal, pp. 111–24. Cranfield, UK: Cranfield Press.

Kelley, J. & Allsopp, D. (1987). Mould growth testing of materials, components and equipment to national and international standards. *Society of Applied Bacteriology, Technical Series* **23**, Oxford, UK: Blackwell Scientific Publications.

Lavappa, K. S. (1978). Trypsin-Giemsa banding procedure for chromosome preparations from cultured mammalian cells. *TCA Manual* **4**, 761–4.

Macy, M. (1978). Identification of cell line species by isoenzyme analysis. *TCA Manual* **4**, 833–6.

Macy, M. (1979). Tests for mycoplasmal contamination of cultured cells as applied at the ATCC. *TCA Manual* **5**, 1151–5.

May, M. C., Grim, E., Wheller, R. M. & West, J. (1982). Determination of residual moisture in freeze-dried viral vaccines: Karl Fischer, gravimetric thermogravimetric methodologies. *J. Biological Standardization* **10**, 249–59.

8

Organisation of resource centres

B. E. KIRSOP and E. J. DASILVA

8.1   Introduction

Individual resource and information centres provide valuable services to biotechnology, but their role can be substantially enhanced if their activities are effectively co-ordinated. This has been recognised in the past, and a number of committees, federations and networks have been set up for this purpose at the national, regional and international levels. Although the origins and composition of existing organisations differ and their geographical locations are widespread, their common purpose is to support and develop the activities of resource and information centres for the benefit of microbiology.

8.2   International organisation
8.2.1   *World Federation for Culture Collections*

There are fewer difficulties in setting up national and regional co-ordinating mechanisms than international systems, and yet one of the first developments in this area was the formation of the World Federation for Culture Collections (WFCC). In 1962 at a Conference on Culture Collections held in Canada it was recommended that the International Association of Microbiological Societies (IAMS) set up a Section on Culture Collections. The Section was established in 1963. Five years later, at an International Conference on Culture Collections in Tokyo, the formation of the WFCC was proposed and an *ad hoc* committee, together with the Section on Culture Collections, drew up statutes which were agreed at a congress in 1970. Following the conversion of the IAMS to Union status, the WFCC is now a federation of the International Union of Microbiological Societies (IUMS) and an interdisciplinary Commission within the International Union of Biological Sciences (IUBS).

173

The principal objective of the WFCC is to establish effective liaison between persons and organisations concerned with culture collections and the users of the collections both in the developed and developing regions of the world. To achieve this objective a structure of committees has been set up covering patents, postal and quarantine regulations, education, endangered collections and publicity.

Committee on Patent Procedures. The activities of the Committee on Patent Procedures have important implications for biotechnology. The procedures for patenting processes involving the use of microorganisms, animal or plant cells or of genetically manipulated organisms are described in Chapter 6. The various patent regulations existing in different parts of the world present a confusing picture to those wishing to take out patents, and professional guidance is essential. A number of organisations such as the World Intellectual Property Organisation (WIPO) are concerned with the rationalisation of the different systems, and the WFCC's patents committee has acted in an advisory capacity to them, providing microbiological input. Members of the Committee have attended WIPO meetings to advise on the implementation of the Budapest Treaty for the International Recognition of the Deposit of Microorganisms for the Purpose of Patent Procedures. Additionally, they have monitored the functioning of the Treaty and provided evidence of difficulties that have arisen in its implementation.

Committee for Quarantine and Postal Regulations. The Committee for Quarantine and Postal Regulations is similarly in close communication with the relevant postal regulatory bodies, such as the International Postal Union, National Postal Departments and the International Air Transport Association (IATA), and has put forward recommendations for the safe transport of infectious and non-infectious biological material (see Chapter 3). Members of the committee have been able to encourage international collaboration in this area by attending appropriate meetings and providing specialist advice in order to establish mechanisms for the safe transport of biological material throughout the world.

Education Committee. The WFCC is aware of the lack of guidance given to students before finishing university training on the support and services available from the microbial resource centres of the world. A similar lack of general awareness exists among many working microbiologists in industry, research and education. Accordingly, the Education Committee of the WFCC has an on-going programme of activities to increase the amount of information on back-up available from culture collections. Projects include the publication of books, preparation of training videos, advisory leaflets, and the organisation of training courses, scientific symposia and international conferences. This present series of source books is part of the programme of the Education Committee, designed to increase the usefulness of culture collections to those working in biotechnology.

Committee for Endangered Collections. The Committee for Endangered Collections is concerned to protect the microbial and cellular genetic resources of the world. Many of the major culture collections suffer from time to time from financial restrictions or from a change of direction in the interests of the host institute. Smaller collections are often transitory in nature and face difficulties on the retirement or relocation of the curator whose special interest the collection represents. The WFCC believes the conservation of these collections is of prime importance if the cultures and the substantial investment in terms of effort and expertise are not to be irretrievably lost. To enable emergency measures to be taken when difficulties arise, the Committee for Endangered Collections has obtained financial backing to set up a fund for the provision of specialist, short-term support to allow the relocation of such collections to alternative laboratories willing and competent to take them over. The services of this committee may be used to provide advice to microbiologists who have developed collections of unique microorganisms during the course of their work, but who may not have the wish or expertise to maintain them in the long term or the resources to supply cultures to others.

Publicity Committee. The WFCC's Publicity Committee plays a major role in the dissemination of information about the activities of the Federation to the microbiological community. It produces a newsletter at regular intervals and is closely involved with all administrative developments. In particular, it plays an important part in the four-yearly WFCC International Conference and in the preparation of posters for scientific conferences. The editor of the newsletter will consider the publication of appropriate material and welcomes information about meetings, publications and topics of general interest to members. Biotechnologists may use the newsletter as a forum for the discussion of issues – possibly controversial – that are of interest to fellow scientists. Typical of subjects that can usefully be discussed in the columns of the newsletter are questions relating to stable nomenclature of microorganisms, the retention of published strain designations, security measures for the release of potentially dangerous cultures to those unqualified to handle them or the rescue of important genetic resources.

Data centres. In addition to the functions of these Committees, and others set up from time to time as the need arises, the WFCC has sponsored and is responsible for the World Data Center of Collections of Cultures of Microorganisms. The Center has pioneered the collection of data of this kind and has been responsible for the publication of three Directories listing the collections and the species they hold. The Center was originally housed in the University of Queensland's Department of Microbiology in Australia, but in 1986, on the retirement of its founder Director, was transferred to the Life Sciences Division at RIKEN, Tokyo, Japan. The WFCC is also co-sponsor with CODATA and IUMS of the international Microbial Strain Data Network (MSDN) set up to provide a referral system to the numerous data centres developing throughout the world listing microbial strain data. These two important activities of the WFCC, set up with international funding, are further discussed in Chapter 2.

The WFCC plays a prime role in the organisation of culture collection activities internationally and has among its membership experts in many areas of microbiology. It exists to serve both the culture collections and their users and may be used as a powerful interdisciplinary organ of communication between biotechnologists and specialists in other areas of microbiology.

8.2.2   *The MIRCEN Network*

The *UNESCO Courier* of July 1975 carried a feature 'On the road to development – a UNESCO network for applied microbiology'. Therein several mechanisms – conferences, training courses and fellowships – were identified. Since then, as a means towards strengthening the world network, several regional and international initiatives have been built in through the establishment of microbiological resources centres (MIRCENs) (see Table 8.1). These are designed:

(1) to provide the infrastructure for the building of a world network which would incorporate regional and interregional functional units geared to the management, distribution and utilisation of the microbial gene pool;

(2) to strengthen efforts relating to the conservation of microorganisms with emphasis on *Rhizobium* gene pools in developing countries with an agrarian base;

(3) to foster the development of new inexpensive technologies that are native to the region;

(4) to promote the applications of microbiology in the strengthening of rural economies;

(5) to serve as focal centres for the training of manpower and the imparting of microbiological knowledge.

Table 8.1. *Microbial resource centres*

Biotechnology MIRCENs

Ain Shams University, Faculty of Agriculture, Shobra–Khaima, Cairo, Arab Republic of Egypt

Applied Research Division, Central American Research Institute for Industry (ICAITI), Ave, La Reforma 4–47 Zone 10, Apdo Postal 1552, Guatemala

Mycology MIRCEN, CAB International Mycological Institute, Ferry Lane, Kew, Surrey TW9 3AF, UK

Department of Bacteriology, Karolinska Institutet, Fack, S–10401 Stockholm, Sweden

Fermentation, Food and Waste Recycling MIRCEN, Thailand Institute of Scientific and Technological Research, 196 Phahonyothin Road, Bangken, Bangkok 9, Thailand

Fermentation Technology MIRCEN, ICME, University of Osaka, Suita-shi 656, Osaka, Japan

Institute for Biotechnological Studies, Research and Development Centre, University of Kent, Canterbury CT2 7TD, UK.

Marine Biotechnology MIRCEN, Department of Microbiology, University of Maryland, College Park Campus, Maryland 207742, USA

Planta Piloto de Procesos Industriales Microbiologicos (PROIMI), Avenida Belgrano y Pasaje Caseros, 4000 S.M. de Tucuman, Argentina

University of Waterloo, Ontario, Canada N2LK 3GI, and University of Guelph, Guelph, Ontario NIG 2WI, Canada

Rhizobium MIRCENs

Cell Culture and Nitrogen-Fixation Laboratory, Room 116, Building 011-A, Barc–West, Beltsville, Maryland 20705, USA

Centre National de Recherches Agronomiques, d'Institut Senégalais de Recherches Agricoles, B.P. 51, Bambey, Senegal

Departments of Soil Sciences and Botany, University of Nairobi, PO Box 30197, Nairobi, Kenya

IPAGRO, Postal 776, 90000 Porto Alegre, Rio Grande do Sul, Brazil

NifTAL Project, College of Tropical Agriculture and Human Resources, University of Hawaii, PO Box 'O', Paia, Hawaii 96779, USA

World Data Center MIRCEN

World Data Center on Collections of Microorganisms, RIKEN, 2–1 Hirosawa, Wako, Saitama 351–01, Japan

The first development in the UNESCO global network of Microbiological Resource Centres, consisting of centres in the developed world and regional networks in the developing countries, was the establishment of the World Data Center (WDC) on Microorganisms (see above and Chapter 2) in Queensland, Australia.

The MIRCEN at the Karolinska Institute, Sweden, in addition to developing microbiological techniques for the identification of microorganisms, has pioneered the organisation of a series of MIRCENET Computer Conferences on biogas production, anaerobic digestion and the bioconversion of lignocellulose. Computer conferencing is a network system that links geographically scattered nodes together through the use of home or office computers to a remote control computer (see Chapter 2).

Apart from attempting to link up the MIRCENs and organising specialised conferences, MIRCENET has other functions listed in Table 8.2.

On the basis of their research and training programmes, the other MIRCENs can be broadly classified as follows.

The Biotechnology MIRCENS. In the area of biotechnology, there are nine MIRCENs in operation (see Table 8.1). These are in Thailand, Egypt, Guatemala, Japan, Argentina, USA, the UK, Canada and Sweden.

In the region of Southeast Asia, the MIRCEN in Bangkok has co-operating laboratories in the Philippines, Indonesia, Singapore, Malaysia and Hong Kong and other institutions in Thailand. It serves the microbiological community in the collection, preservation, identification and distribution of microbial germplasm, and in the promotion of research and training activities directed towards the needs of the region.

In the region of the Arab States, the MIRCEN at Ain-Shams University, Cairo, promotes research and training courses on the conservation

Table 8.2. *Functions of MIRCENET*

To help initiate closed computer conferences under defined keys such as microbiology, biological nitrogen fixation, biogas, networking in culture collections.

To act as an information source for meetings, reviews, identification services, etc.

To provide a platform for discussions on MIRCEN network activities.

To provide print-outs and records of MIRCENET entries.

of microbial cultures and biotechnologies of interest to the region. Through its co-operating MIRCEN laboratory at the University of Khartoum, the MIRCEN has contributed to the establishment of a culture collection in Sudan specialising in fungal taxonomy. The co-operating MIRCEN laboratory at the Institute Agronomique et Veterinaire Hassan II, Rabat, has made commendable progress through projects using different species of yeasts and rhizobia.

In the region of Central America and the Caribbean, the MIRCEN (co-operating laboratories in Chile, Columbia, Costa Rica, Dominican Republic, Ecuador, El Salvador, Honduras, Jamaica, Mexico, Nicaragua, Peru, Venezuela) has, in co-operation with the Organization of American States, the InterAmerican Development Bank and several other prestigious agencies, pioneered the applications of microbiology, process engineering and fermentation technology in several member states of Central America and the Caribbean. It has set up joint collaborative research projects, the exchange of technical personnel, regional training programmes and the dissemination of scientific information among network institutions.

The South American Biotechnology MIRCEN located at Tucuman, Argentina, is comprised of a regional network with co-operating laboratories in Brazil, Chile, Bolivia and Peru. It has similar goals as the Biotechnology MIRCEN for Central America and the Caribbean.

The MIRCENs in the industrialised societies function as a bridge with those in the developing countries. In such manner, increased co-operation is promoted between the developed and developing countries. Furthermore, a basic structure is set up for eventual twinning at a later date. For example, the Guelph Waterloo MIRCEN, Canada, with its expertise at the University of Waterloo in biomass conversion technology, microbial biomass protein production and bioreactor design, is of immense benefit to the work of the MIRCENs at Cairo, Guatemala and Tucuman.

In a similar manner, the MIRCEN at Bangkok has several collaborative research projects with that at the International Centre of Co-operative Research in Biotechnology, Osaka, Japan. This centre conducts the annual UNESCO International Postgraduate University Course on Microbiology (of 12 months' duration). It also functions as the Japanese point-of-contact for the Southeast Asian regional network of microbiology in the UNESCO Programme for Regional Co-operation in the basic sciences.

In the UK there is a MIRCEN network centred upon the Institute for

Biotechnological Studies (IBS). In common with other Microbiological Resource Centres, the aims of the UK MIRCEN are to promote the utilisation of the microbial gene pool, to promote applied microbiology and biotechnology in the developing countries and to provide a centre for training and advice.

The intergovernmental CAB International Mycological Institute (CMI) is the MIRCEN for mycology co-operating with all others in the field world-wide. This organisation, together with the Institute of Horticultural Research (IHR) are the first two organisations to collaborate with the MIRCEN Network, whilst continuing their own activities in mycological and biodeterioration studies, and microbiological pest control, mycorrhizas and mushroom technology respectively.

In keeping with the new trends of the expanding frontiers of biotechnological research, a Marine Biotechnology MIRCEN has been established at the University of Maryland. Work presently underway includes the fundamental elucidation of the evolution of genes and the flow of genes through populations in the marine environment. One collaborative study underway is with the Chinese University of Hong Kong and the Shandong College of Oceanography, Qingdao, China.

The Biological Nitrogen Fixation (BNF) MIRCENS. In the quest for more food for their increasing populations, several developing nations have been expanding their agricultural lands into areas which are marginally capable of sustaining productivity and invariably limited by the availability of nitrogen fertilizer.

In interaction with other international programmes, modest schemes for the development of biofertilizers or *Rhizobium* inoculant material, particularly in legume-crop areas of the developing countries, are already operating through the MIRCENs on a level of regional co-operation in Latin America, East Africa and Southeast Asia and the Pacific.

In the area of biological nitrogen five MIRCENs are already operating (see Table 8.1).

The broad responsibilities of these MIRCENs include collection, identification, maintenance, testing and distribution of rhizobial cultures compatible with crops of the regions. Deployment of local rhizobia inoculant technology and promotion of research are other activities. Advice and guidance are provided in the region to individuals and institutions engaged in rhizobiology research.

The BNF MIRCENs play a valuable role in maintaining and distributing efficient cultures of *Rhizobium*. Nearly 4000 strains are maintained in the MIRCEN collections and about 1750 have been distributed to other organisations (Table 8.3).

The MIRCEN network is founded on the principle of self-help and mutual co-operation. It concentrates on existing facilities and resources and provides an organisational structure which allows each institution to collaborate as best it can through the following:

(1) an exchange of research workers between national and regional institutions;

(2) small grants to individual research projects or workers for

Table 8.3. *Culture collection services of Biological Nitrogen Fixation (BNF) MIRCENs*

Holdings of Rhizobium *culture collections*

MIRCEN	Number of strains held
Bambey	50
Beltsville	938
Hawaii	2000
Nairobi	208
Porto Alegre	650
Total	3846

Cultures distributed by Rhizobium *MIRCENs*

MIRCEN	Number of cultures	Countries of recipient institutions
Bambey	8	Gambia, Mali, Yemen
Beltsville	508	Zimbabwe, Nigeria, Yugoslavia, India, Spain, Vietnam, Ireland, UK, Malaysia, Italy, Canada, South Africa, Senegal, Egypt, Poland, Argentina, Turkey, W. Germany, Austria, Australia, New Zealand
Hawaii	200	Global
Nairobi	95	Uganda, Malawi, Tanzania, Mauritius, Sudan, Congo, Zaire, Rwanda
Porto Alegre	943	Argentina, Chile, Bolivia, Uruguay, Peru, Ecuador, Columbia, Venezuela, El Salvador, Dominican Rep., Mexico, USA, Trinidad, Brazil

acquisition of supplies, spare parts for equipment, or small-scale equipment;

(3) participation of senior scientists in specialised symposia in the technically advanced countries in the vicinity of each of the regions;

(4) organisation of short-term intensive training courses and specialised in-depth sub-national or national meetings;

(5) production of a newsletter functioning as an outlet for the exchange of research news, publication of research findings and as an attraction for potential participating laboratories.

The MIRCENs play a catalytic role in breaching the barrier of geographical isolation and advancing the frontiers of contemporary research in biotechnology through the production of newsletter bulletins, culture collection catalogues and research papers. The publication of *MIRCEN News* annually, the development of MIRCENET, and the UNESCO *MIRCEN Journal of Applied Microbiology and Biotechnology* are indications of the gradual emergence of competence and capability of the MIRCENs, and the services they provide on a regional and inter-regional basis.

8.3   Regional organisation

Transnational co ordinating mechanisms are being set up throughout the world to bring regional cohesion to culture collection activities and benefit to both the resource centres themselves and their users. Some have been established as committees by culture collections; others have originated as data centres with the secondary effect of stimulating closer working collaboration between the contributing culture collections. They may be contacted for information about microbiological resources, services and general advice.

8.3.1   *European Culture Collections' Organisation (ECCO)*

In 1981, at an international conference in Brno, Czechoslovakia, curators of European service culture collections present agreed that a mechanism should be set up to enable meetings to take place on an annual basis for the exchange of ideas and the discussion of common problems. In 1982 the first meeting of ECCO took place at the Deutsche Sammlung von Mikroorganismen, Göttingen, FRG, and since then meetings have been held in France, the UK and Czechoslovakia. Membership has increased steadily as new culture collections are formed or developed to provide a national service. Membership is restricted to collections that provide a service on demand and without restriction, that have as a normal part of their duty the acceptance of cultures, that issue from time to time a list of their holdings, and that are in a country with a microbiological society belonging to the Federation of European Microbiological Societies. It was felt that these collections have interests and problems in common that are not shared by research or teaching collections.

Apart from the exchange of scientific information relating to such topics as taxonomy, identification and preservation procedures, much benefit has been derived from discussions on new developments within culture collections such as acceptance of International Depositary Authority Status (see Chapter 6) or the development of computerised systems for the storage, searching and dissemination of culture information (see Chapter 2). In addition, the opportunity to meet on a regular basis has enabled collaborative programmes to be set up between collections from different countries.

ECCO members have become aware that the services available from the collections are not fully exploited by users. To remedy this they have combined to produce publicity material in the form of brochures, leaflets and scientific posters and plan to set up a permanent information centre in the future. Information about the holdings and services of ECCO collections is available from the Organisation's Officers (see Chapter 2) or the Secretary of FEMS.

8.3.2 *Regional database systems*

A number of co-ordinating mechanisms based on information centres have been set up, primarily to establish data banks for regional access. These are described in greater detail in Chapter 2.

Some, such as the Tropical Data Base in Brazil, the Microbial Information Network in Europe (MINE) and the Nordic Register, have been developed initially to provide a centre for information on the culture collections themselves, their services and their holdings. Others, such as the Microbial Culture Information Service (MiCIS), have been set up in areas with well-established culture collection systems with the purpose of providing on-line strain database for searching. Both these kinds of data centres have the secondary effect of encouraging collaboration between culture collections so that the best possible system develops and minimum duplication of effort takes place.

The proliferation of microbial data centres world-wide reflects the growing need for information on biological materials. It has also led to problems in identifying the most appropriate point of contact for specific information. To overcome this an international system has been established (Microbial Strain Data Network, MSDN), to act as a referral system and communications network to databases able to answer specific enquiries on strain properties (Chapter 2). It seems certain that other systems will be established, and the function of the MSDN thus becomes of increasing importance as the first point of enquiry, directing those seeking information on strain properties to appropriate centres.

8.4   National federations/committees

The following countries have established federations or committees for the co-ordination of culture collection activities.

Australia
Canada
China
Czechoslovakia
Japan
Korea
New Zealand
Turkey
United Kingdom
United States of America

Information about them and their activities may be obtained through culture collections or microbiological societies within the country or through the World Data Center and the Microbial Strain Data Network (see Chapter 2). Most of these organisations produce newsletters from time to time and further information may be obtained through these publications.

Some of the organisations are *for* culture collections, others are *of* culture collections and the difference between the two categories is significant. Those that are *of* culture collections exist primarily to co-ordinate culture collection activities within the countries (produce common catalogues, rationalise holdings, stabilise funding) and are generally termed Committees rather than Federations; those that are *for* culture collections have as their prime function the promotion of communication between the collections and their users in industry, research and education. The activities of the latter category concentrate more on scientific meetings, workshops and training courses, and the membership includes any microbiologists with an interest in culture collection activities, whether they are working in culture collections or not. The executive boards are deliberately formed of people both from culture collections and from research or teaching laboratories and industry, providing a cross fertilisation of interests, whereas with organisations set up *for* culture collections the officers and members are drawn from the collections only. The impact of biotechnological input to the Federations has played a valuable part in the development of microbial resource centres to meet the growing needs of industry in this area.

A number of international organisations exist for the co-ordination of activities within different microbiological disciplines, and information about them can be obtained from the International Council for Scientific Unions (Table 8.4). Information on biotechnology is disseminated through the different associations listed in Table 8.5. All these organisations recognise the need for an effective network of microbial resource centres and are active in support of their development.

8.5 Future developments

Developments in biotechnology have coincided with extensive advances in computer technology, and throughout the world culture collections have taken advantage of the latter to respond to the increasing demands of the former. It is clear from Chapter 2 that data held in the microbial resource centres are increasingly computerised and it is evident that the biotechnology community can better be served by co-ordination of these activities. The World Data Center for Collections of Cultures of Microorganisms and the Microbial Strain Data Network are important examples of international collaboration in this area, leading to on-line databases and information network systems. The imaginative and successful MIRCEN network will continue to be instrumental in

Table 8.4. *International scientific organisations*

ICSU	International Council of Scientific Unions 51 Boulevard de Montmorency 75016 Paris France Telephone: 45.25.03.29 Telex: ICSU 630553 F
IUBS	International Union of Biological Sciences
IUMS	International Union of Microbiological Societies
ICRO	(International Cell Research Organisation) Panel on Applied Microbiology and Biotechnology

encouraging the establishment and development of culture collection activities in the developing world and linking them to those in industrial nations.

Computers will be used increasingly for computer conferencing and electronic mail, leading to greater communication between the collections. This in turn should lead to greater collaborative research and joint service activities and will minimise unnecessary duplication of effort, consistant with national requirements.

In spite of rapid developments in communication systems, the need for the presence of culture collections in all regions of the world will

Table 8.5. *Biotechnology associations*

AABB	Association for the Advancement of British Biotechnology 1 Queen's Gate London SW1H 9BT UK
ABA	Australian Biotechnical Association 1 Lorraine Street Hampton Victoria 3188 Australia
ABC	Association of Biotechnology Companies 1220 L Street NW Suite 615 Washington, DC 20005 USA
ADEBIO	Association de Biotechnologie 3 rue Massenet 77300 Fontainebleau France
BIDEC	c/o Japan Association of Industrial Fermentation 20–5 Shinbashi 5-chome Minato-ku Tokyo 105 Japan
IBA	Industrial Biotechnology Association 2115 East Jefferson Street Rockville Maryland 20852 USA
IBAC	Industrial Biotechnology Association of Canada Lava University Cité Universitaire Quebec G1K 7P4 Canada

remain because of specialised local needs, regional regulatory requirements, such as those for postal and quarantine purposes, or currency or
language reasons. Duplication of important holdings and services is
necessary, but can be reduced to an acceptable level by collaborative
efforts on the part of individual scientists in the resource centres, the
setting up of organisations to co-ordinate their activities and the use of
computers and electronic networking to facilitate communication. A
basic core of collaborative mechanisms already exists and can be
extended to cover regions of the world or specialist areas of activity not
yet co-ordinated internationally.

APPENDIX: MEDIA

A. H. S. ONIONS and J. I. PITT

This Appendix gives details of some commonly used media for the culture of filamentous fungi, mainly from Smith & Onions (1983). Special media have been developed for particular fungi, for isolation, identification and maintenance. For further details of these see the compilations of mycological media given in Booth (1971b), Constantinescu (1974), Malloch (1981), Pitt & Hocking (1985) for the food industry, and Stevens (1974). The recipes are arranged in alphabetical order. Preparation methods follow the usual procedure unless otherwise stated. Tables A.1 and A.2 list the media and their uses.

Carnation leaf agar (Fisher et al., 1980)   CLA

Healthy carnation leaves (free from fungicide or insecticide residues) are cut into 5-mm pieces immediately after collection and then dried at about 70 °C for 2 hours. They are then sterilised by either gamma irradiation or propylene oxide vapour taking necessary safety precautions.

The sterile leaf pieces are placed aseptically into Petri dishes and molten 2% TWA (see below) added; one piece per 2 ml agar is adequate.

Used for growth of *Fusarium*.

Cornmeal (maize) agar   CMA
Maize   30 g
Agar   20 g
Water   1 litre

Peptone (20 g) is also sometimes added.
Place the maize in the water (if meal is not available break up 30–35 g of grain and pass through a coffee mill), heat in a water bath or double

saucepan until boiling, stirring for 1 hour. Filter the decoction through muslin, add the agar, and heat until it is dissolved.

Autoclave for 15 minutes at 121 °C.

Table A.1. *Uses of media*

Medium	Use
CLA	*Fusarium*
CMA	Dematiaceous fungi and general use
Cz	*Aspergillus* and *Penicillium*
Cz20S	*Eurotium*
CYA	*Aspergillus* and *Penicillium*
CY20S	*Eurotium*
DG18	Xerophilic fungi
MA	General purpose
M/20, M/40, M/60, MSalt	Xerophilic fungi
MCz	*Aspergillus* and *Penicillium*
MEA	General purpose
OA	Cellulolytic fungi, *Phytophthora* and *Pythium*
PCA	Starvation medium to induce sporulation
PDA	General purpose, tends to encourage mycelial growth – rather rich
PSA	*Fusarium*
TWA	Starvation medium, good as support for solid substrates, filter paper, straw and vegetable pieces
YPSS	Thermophilic fungi

Note: For general purposes use CMA, Cz, MCz, MA or PDA.

Table A.2. *Media for growth and identification of specific genera*

Genera	Media
Aspergillus and *Penicillium*	Cz, CYA, MA, MCz
Dematiaceous fungi	CMA, PCA, TWA+
Eurotium	Cz20S, CY20S
Fusarium	CLA, PSA
Phytophthora and *Pythium*	OA
Thermophilic fungi	YPSS
Xerophilic fungi	Cz20S, CY20S, DE18, M/20–60, MSalt

Czapek agar   Cz

Stock solution A   50 ml
Stock solution B   50 ml
Distilled water   900 ml
Sucrose   30 g
Agar   20 g

Stock solution A

Sodium nitrate $NaNO_3$ 40 g
Potassium chloride KCl   10 g
Magnesium sulphate $MgSO_4 \cdot 7H_2O$ 10 g
Ferrous sulphate $FeSO_4 \cdot 7H_2O$ 0.2 g
Dissolve in 1 litre distilled water and store in a refrigerator.

Stock solution B

Dipotassium hydrogen orthophosphate K_2HPO_4 20 g
Dissolve in 1 litre distilled water and store in a refrigerator.
To each mixed litre of medium then add 1.0 ml of both (a) and (b):
   (a) Zinc sulphate $ZnSO_4 \cdot 7H_2O$ 1.0 g in 100 ml water
   (b) Copper sulphate $CuSO_4 \cdot 5H_2O$ 0.5 g in 100 ml water
Autoclave for 20 minutes at 121 °C.
20% sucrose may be added for growth of *Eurotium* (**Cz20S**)

Czapek yeast autolysate   CYA

Dipotassium hydrogen orthophosphate K_2HPO_4 1 g
Czapek concentrate (*see below*)   10 ml
Yeast extract   5 g
Sucrose   30 g
Agar   15 g
Distilled water   1 litre

Czapek concentrate

Sodium nitrate $NaNO_3$ 30 g
Potassium chloride KCl   5 g
Magnesium sulphate $MgSO_4 \cdot 7H_2O$ 5 g
Ferrous sulphate $FeSO_4 \cdot 7H_2O$ 0.1 g
Water   100 ml
Addition of trace elements $ZnSO_4 \cdot 7H_2O$ (0.1 g) and $CuSO_4 \cdot 5H_2O$ (0.05 g) improve this medium.
Czapek concentrate will keep indefinitely without sterilisation.
Autoclave for 15 minutes at 121 °C.
20% sucrose may be added for growth of *Eurotium* (**CY20S**).

Dichloran 18% glycerol agar   DG18

Glucose	10 g
Peptone	5 g
Potassium dihydrogen orthophosphate KH_2PO_4	1 g
Magnesium sulphate $MgSO_4 \cdot 7H_2O$	0.5 g
Glycerol	220 g
Agar	15 g
Distilled water	1 litre
Dichloran	2 mg (0.2% w/v in ethanol, 1 ml)
Chloramphenicol	100 mg

Autoclave for 15 minutes at 121 °C.

Malt Czapek agar   MCz

Czapek solution A (see above)	50 ml
Czapek solution B (see above)	50 ml
Sucrose	30 g
Malt extract	40 g
Agar	20 g
Distilled water	900 ml

Dissolve malt extract and agar in water. Add Czapek stock solutions A and B and sucrose and heat gently in a water bath or double saucepan until dissolved. Adjust pH to between 4 and 5. Autoclave for 20 minutes at 121 °C.

Malt extract agar   MEA

Malt extract (powdered, Difco or Oxoid)	20 g
Peptone (bacteriological)	1 g
Glucose	20 g
Agar	15 g
Distilled water	1 litre

Autoclave for 15 minutes at 121 °C.

Malt extract agar (2%)   MA

Malt extract (good quality)	20 g
Agar	20 g
Water	1 litre

Dissolve malt extract in water, add agar and dissolve. Autoclave for 20 minutes at 121 °C.

The brand of malt extract appears to be an important factor with this medium. Do not filter after adding the malt. The pH will be between 3 and 4, and should be adjusted to 6.5 with sodium hydroxide solution.

The addition of 20, 40, and 60% sucrose (M/20, M/40, M/60) makes this an appropriate medium for xerophilic fungi. 50% glucose (MY50G) can also be used for fastidious xerophiles. 10% NaCl may also be added (Malt Salt).

Oat agar   OA

Powdered oatmeal	30 g
Japanese Kobe or other pure agar	20 g
Water	1 litre

Gradually heat oatmeal and water to boiling in a water bath or double saucepan, stirring occasionally. Simmer for 1 hour. Pass through muslin, make up to 1 litre, add agar, and heat until dissolved. Autoclave for 20 minutes at 121 °C.

0.5 ml wheatgerm oil may be added but requires thorough mixing. This is useful for cultures of *Phytophthora*.

Potato media

New potatoes should be avoided when making PCA, PDA, and PSA.

Potato carrot agar   PCA

Grated potato	20 g
Grated carrot	20 g
Agar	20 g
Tap water	1 litre

Wash, peel and grate potatoes and carrots as required. Boil vegetables for about 1 hour in 1 litre of tap water. Drain through a fine sieve and add agar. Heat in a water bath or double saucepan until the agar is dissolved. Autoclave for 20 minutes at 121 °C.

Potato dextrose agar   PDA

Potato cubes	200 g
Dextrose	15 or 20 g
Agar	20 g
Water	1 litre

Scrub potatoes clean, and cut up without peeling into 12-mm cubes. Rinse cubes rapidly under a running tap, and place in 1 litre of water in a saucepan. Boil until potatoes are soft (about 1 hour); filter through a sieve and squeeze through as much pulp as possible. Alternatively, blend, in which case a thicker, opaque extract is produced. Add agar, and heat until it dissolves. Add dextrose and stir until dissolved. Make up to 1 litre. Keep stirring while dispensing to ensure even distribution of solids.

Autoclave for 20 minutes at 121 °C.

Potato sucrose agar   PSA

Potato water (see below)	500 ml
Sucrose	20 g
Agar	20 g
Distilled water	500 ml

Potato water

Potatoes   1.8 kg
Water   4.5 litres
Suspend peeled diced potatoes in double cheesecloth in water and boil until potatoes almost cooked (about 8 minutes). Decant water.

Heat in water bath or double saucepan until agar is dissolved. Adjust pH to 6.5 with calcium carbonate if necessary.
Autoclave for 15 minutes at 121 °C.

Tap water agar   TWA

Tap water   1 litre
Agar   15 g

Autoclave for 20 minutes at 121 °C.
Sterilised pieces of filter paper or straw can be added to encourage the sporulation of cellulolytic fungi.

Yeast phosphate soluble starch   YPSS

'Difco yeast extract'	4 g
Soluble starch	15 g
Dipotassium hydrogen ortho- phosphate K_2HPO_4	1 g
Magnesium sulphate $MgSO_4 \cdot 7HO_2$	0.5 g
Agar	20 g
Water	1 litre

Autoclave for 15 minutes at 121 °C.

Isolation media

Pitt & Hocking (1985) include several inhibitory media containing various quantities and combinations of rosebengal, dichloran and chloramphenicol primarily for use in the food spoilage industry for isolation purposes, e.g. RBC, DRBC, DCPA. Antibiotics such as penicillin and streptomycin or lowering of the pH as in Czapek's agar can also be used for this purpose (Onions *et al.*, 1981).

REFERENCES

In re Abitibi (1982). *Canadian Patent Reporter* **62**, 81.

Alexander, M., Daggett, P.-M., Gherna, R., Jong, S. C., Simione, F. & Hatt, H. (1980). *American Type Culture Collection Methods* I. *Laboratory Manual on Preservation Freezing and Freeze Drying*. Rockville, Maryland: American Type Culture Collection.

Anderson, C. & Solomons, G. L. (1984). Primary metabolism and biomass production from *Fusarium*. In *The Applied Mycology of Fusarium*, ed. M. O. Moss & J. E. Smith, pp. 231–50. Cambridge University Press.

Ardsell, J. N. van, Kwok, S., Schweickart, V. L., Ladner, M. B., Gelfand, D. H. & Innis, M. A. (1987). Cloning, characterization, and expression in *Saccharomyces cerevisiae* of endoglucanase I from *Trichoderma reesei*. *Bio/ Technology* **5**, 60–4.

Batra, L. K. & Millner, P. D. (1976). Asian fermented foods and beverages. *Developments in Industrial Microbiology* **17**, 117–28.

Beier, F. K., Crespi, R. S. & Straus, J. (1985). *Biotechnology and Patent Protection: An International Review*. Paris: Organization for Economic Co-operation and Development.

Bennett, J. W. (1985). Molds, manufacturing and molecular genetics. In *Molecular Genetics of Filamentous Fungi*, ed. W. E. Timbalake, pp. 345–66. New York: Alan Liss.

Bennett, J. W. & Lasure, L. L. (1985). *Genetic Manipulations in Fungi*. Orlando: Academic Press.

Birch, G. G., Parker, K. J. & Worgan, J. T. (ed.) (1976). *Food From Waste*. London: Applied Science Publishers.

Booth, C. (1971a). *The Genus Fusarium*. Kew: Commonwealth Mycological Institute.

Booth, C. (ed.) (1971b). *Methods in Microbiology*, vol. 4. London & New York: Academic Press.

Bridge, P. D. (1985). An evaluation of some physiological and biochemical methods as an aid to the characterization of species of *Penicillium* subsection *Fasciculata*. *Journal of General Microbiology* **131**, 1887–95.

Bridge, P. D. & Hawksworth, D. L. (1984). The API ZYM enzyme testing system as an aid to the rapid identification of *Penicillium* isolates. *Microbiological Sciences* 1, 232–4.

Bridge, P. D. & Hawksworth, D. L. (1985). Biochemical tests as an aid to the identification of *Monascus* species. *Letters in Applied Microbiology* 1, 25–9.

Bridge, P. D., Hawksworth, D. L., Kozakiewicz, Z., Onions, A. H. S., Paterson, R. R. M. & Sackin, M. J. (1986). An integrated approach to *Penicillium* systematics. In *Advances in Penicillium and Aspergillus Systematics*, ed. R. A. Samson & J. I. Pitt, pp. 281–309. New York & London: Plenum Press.

Budapest Treaty (1981). *Budapest Treaty on the International Recognition of the Deposit of Microorganisms for the Purposes of Patent Procedure* 1977 and *Regulations* 1981. Geneva: World Intellectual Property Organization.

Buell, C. B. & Weston, W. H. (1947). Application of the mineral oil conservation method to maintaining collections of fungus cultures. *American Journal of Botany* 34, 555–61.

Butterfield, W., Jong, S. C. & Alexander, M. J. (1974). Preservation of living fungi pathogenic for man and animals. *Canadian Journal of Microbiology* 20, 1665–73.

Cannon, P. F. (1986). Name changes in fungi of microbiological, industrial and medical importance, I–II. *Microbiological Sciences* 3, 168–71, 285–7.

Carmichael, J. W. (1962). Viability of mould cultures stored at $-20\,°C$. *Mycologia* 54, 432–6.

Casida, L. E. (1964). *Industrial Microbiology*. London: John Wiley & Sons.

Chang, S. T. & Hayes, W. A. (ed.) (1978). *The Biology and Cultivation of Edible Mushrooms*. London & New York: Academic Press.

Charles, M. (1985). Fermentation scale up: problems and possibilities. *Trends in Biotechnology* 3, 134–9.

Cole, R. J. & Cox, R. H. (1981). *Handbook of Toxic Fungal Metabolites*. London: Academic Press.

Constantinescu, O. (1974). *Metode si tehnici in Micologie*. Bucurest: Editura Ceres.

Convention (1980). *Convention on the Grant of European Patents*. (European Patent Convention) 1973 with 1980 amendments. Munich: European Patent Organization.

Crespi, R. S. (1982). *Patenting in the Biological Sciences*. Chichester: John Wiley & Sons.

Crespi, R. S. (1985). Patent protection in biotechnology: questions, answers and observations. In *Biotechnology and Patent Protection*, ed. F. K. Beier, R. S. Crespi & J. Straus, pp. 36–85. Paris: Organization for Economic Co-operation and Development.

Cruickshank, R. H. & Pitt, J. I. (1987). The zymogram technique: isoenzyme patterns as an aid in *Penicillium* classification. *Microbiological Sciences* 4, 14–17.

Crush, J. R. & Pattison, A. C. (1975). Preliminary results on the production of vesicular-arbuscular mycorrhizal inoculum by freeze drying. In

Endomycorrhizas, ed. F. E. Sanders, B. Mosse & P. B. Tinker, pp. 485–509. London: Academic Press.

Dahmen, H., Staub, T. & Schwinn, F. T. (1983). Technique for long-term preservation of phytopathogenic fungi in liquid nitrogen. *Phytopathology* **73**, 241–6.

In re Diamond & Chakrabarty (1980). *US Patents Quarterly* **206**, 193.

Eggins, H. O. W. & Allsopp, D. (1975). Biodeterioration and biodegradation by fungi. In *The Filamentous Fungi*, ed. J. E. Smith & D. R. Berry, pp. 301–20. London: Edward Arnold.

Ellis, J. J. (1979). Preserving fungus strains in sterile water. *Mycologia* **71**, 1072–5.

Ellis, J. J. & Roberson, J. A. (1968). Viability of fungus cultures preserved by lyophilization. *Mycologia* **60**, 399–405.

Emmons, C. W., Binford, C. H., Utz, J. P. & Kwon-Chung, K. J. (1977). *Medical Mycology*, 3rd edn. Philadelphia: Lea & Febiger.

European Culture Collections' Organisation (1984). *Services in Microbiology*. Norwich: European Culture Collection Curators' Organization.

Evans, E. G. V. & Gentles, J. C. (1985). *Essentials of Medical Mycology*. Edinburgh: Churchill Livingstone.

Fisher, N. L., Burgess, L. W., Toussoun, T. A. & Nelson, P. E. (1980). Carnation leaves as a substrate and for preserving cultures of *Fusarium* species. *Phytopathology* **72**, 151–3.

Frisvad, J. C. (1981). Physiological criteria and mycotoxin production as aids in identification of common asymmetric Penicillia. *Applied and Environmental Microbiology* **46**, 1301–10.

Goldie Smith, E. K. (1956). Maintenance of stock cultures of aquatic fungi. *Journal of the Elisha-Mitchell-Scientific Society* **72**, 158–66.

Haskins, R. H. (1957). Factors affecting survival of lyophilised fungal spores and cells. *Canadian Journal of Microbiology* **3**, 477–85.

Hawksworth, D. L. (1985*a*). Fungus culture collections as a biotechnological resource. *Biotechnology and Genetic Engineering Reviews* **3**, 417–53.

Hawksworth, D. L. (1985*b*). The Commonwealth Mycological Institute (Kew). *Biologist* **32**, 7–12.

Hawksworth, D. L., Sutton, B. C. & Ainsworth, G. C. (1983). *Ainsworth & Bisby's Dictionary of the Fungi*, 7th edn, 445 pp. Kew: Commonwealth Mycological Institute.

Heckly, R. J. (1978). Preservation of microorganisms. *Advances in Applied Microbiology* **24**, 1–53.

Hesseltine, C. W. (1965). A millenium of fungi, food and fermentation. *Mycologia* **57**, 149–97.

Hewitt, W. B. & Chiarappa, L. (eds.) (1977). *Plant Health and Quarantine in International Transfer of Genetic Resources*, 346 pp. Cleveland, Ohio: CRC Press.

Hoog, G. S. de (ed.) (1979). *Centraalbureau voor Schimmelcultures. 75 Years Culture Collection*. Baarn & Delft: Centraalbureau voor Schimmelcultures.

Howard, D. H. (ed.) (1983–5). *Fungi Pathogenic for Humans and Animals*, 3 vols. New York & Basel: Marcel Dekker.

Hwang, S.-W. (1966). Long term preservation of fungus cultures with liquid nitrogen refrigeration. *Applied Microbiology* **14**, 784–8.

Hwang, S.-W. & Howells, A. (1968). Investigation of ultra-low temperature for fungal cultures. II. Cryoprotection afforded by glycerol and dimethyl sulphoxide to 8 selected fungal cultures. *Mycologia* **60**, 622–6.

International Convention (1978). *International Convention for the Protection of new Varieties of Plants* 1961, revised 1972. Geneva: World Intellectual Property Organization.

Johnston, A. & Booth, C. (1983). *Plant Pathologist's Pocketbook*, 2nd edn. Kew: Commonwealth Mycological Institute.

Jong, S. C. & Atkins, W. B. (1985). Conservation, collection and distribution of cultures. In *Fungi Pathogenic to Humans and Animals*, ed. D. H. Howard, vol. B (2), pp. 153–94. New York & Basel: Marcel Dekker.

Jong, S. C. & Davis, E. E. (1978). Conservation of reference strains of *Fusarium* in pure culture. *Mycopothologia* **66**, 153–9.

Jong, S. C., Levy, A. & Stevenson, R. E. (1984). Life expectancy of freeze dried fungus cultures stored at 4 °C. In *Proceedings of the Fourth International Conference on Culture Collections*, ed. M. Kocur & E. DaSilva, pp. 125–36. London: World Federation for Culture Collections.

Joshi, L. M., Wilcoxson, R. D., Gera, S. D. & Chatterjee, S. C. (1974). Preservation of fungal cultures in liquid air. *Indian Journal of Experimental Biology* **12**, 598–9.

Kirsop, B. E. (1980). *The Stability of Industrial Organisms*. Kew: CAB International Mycological Institute.

Kirsop, B. E. & Snell, J. J. S. (1984). *Maintenance of Microorganisms*. London: London: Academic Press.

Kramer, C. L. & Mix, A. J. (1957). Deep freeze storage of fungus cultures. *Transactions of the Kansas Academy of Science* **60**, 58–64.

Kurtzman, C. P. (1986). The ARS Culture Collection: present status and new directions. *Enzyme Microbial Technology* **8**, 328–33.

In re Lundak (1985). *US Patents Quarterly* **227**, 90.

McGinnis, M. R. (1980). *Laboratory Handbook of Medical Mycology*. New York: Academic Press.

McGowan, V. F. & Skerman, V. B. D. (1982). *World Directory of Collections of Cultures of Microorganisms*, 2nd edn. St Lucia: World Data Center.

Malloch, D. (1981). *Moulds. Their isolation, cultivation, and identification*. University of Toronto Press.

Moreau, C. (1979). *Moulds, Toxins and Food*. Chichester: John Wiley & Son.

Onions, A. H. S. (1977). Storage of fungi by mineral oil and silica gel for use in the collection with limited resources. In *Proceedings of the Second International Conference on Culture Collections, University of Queensland, World Federation for Culture Collections*, ed. A. F. Pestana de Castro, E. J. DaSilva, V. B. D. Skerman, W. W. Leveritt, pp. 104–13. UNESCO/UNEP/ICRO/WDC, Australia.

Onions, A. H. S. (1983). Preservation of fungi. In *The Filamentous Fungi 4, Fungal Technology*, ed. J. E. Smith, D. R. Berry & B. Kristiansen, pp. 373–90. London: Edward Arnold.

Onions, A. H. S., Allsopp, D. & Eggins, H. O. W. (1981). *Smith's Introduction to Industrial Mycology*, 7th edn. London: Edward Arnold.

Paterson, R. R. M. (1986). Standardized one- and two-dimensional thin-layer chromatographic methods for the identification of secondary metabolites in *Penicillium* and other fungi. *Journal of Chromatography* **368**, 249–64.

Pitt, J. I. (1980) ('1979'). *The Genus Penicillium and its Teleomorphic States Eupenicillium and Talaromyces*. London: Academic Press.

Pitt, J. I. & Hocking, A. D. (1985). *Fungi and Food Spoilage*. Sydney: Academic Press.

Raper, K. B. & Alexander, D. F. (1945). Preservation of molds by lyophil process. *Mycologia* **37**, 499–525.

Rogan, M. & Terry, C. S. (1973). Accelerated stability testing of freeze dried separation of *Penicillium chrysogenum* W1554-1255. *Journal of Pharmacy and Pharmocology* **25**, 134–5.

Rogosa, M., Krichevsky, M. I. & Colwell, R. R. (1986). *Coding Microbiological Data for Computers*. New York: Springer-Verlag.

Rowe, T. W. G. & Snowman, J. W. (1978). *Edwards Freeze Drying Handbook*. Crawley: Edwards High Vacuum.

Ruffles, G. (1986). Patents and the biologist. *Biologist* **33**, 5–10.

Samson, R. A. & Pitt, J. I. (ed.) (1986) ('1985'). *Advances in Penicillium and Aspergillus Systematics*. NATO Advanced Study Institute Series A 102. New York & London: Plenum Publishing.

Sigler, L. & Hawksworth, D. L. (1987). Code of practice for systematic mycologists. *Microbiological Sciences* **4**, 83–6; *Mycopathologia* **99**, 3–7; *Mycologist* **21**, 101–5.

Smith, D. (1983). Cryoprotectants and the cryopreservation of fungi. *Transactions of the British Mycological Society* **80**, 360–3.

Smith, D. (1986). The evaluation and development of techniques for the preservation of living filamentous fungi. PhD. thesis, University of London.

Smith, D., Morris, G. J. & Coulson, G. E. (1986). A comparative study of the morphology of hyphae of *Penicillium expansum* and *Phytophthora nicotianae* during freezing and viability on thawing. *Journal of General Microbiology* **132**, 2013–21.

Smith, D. & Onions, A. H. S. (1983). *The Preservation and Maintenance of Living Fungi*. Kew: Commonwealth Mycological Institute.

Smith, J. E. (1981). *Biotechnology*. London: Edward Arnold.

Steinkraus, K. K. (ed.) (1983). *Handbook of Indigenous Fermented Foods*. New York & Basel: Marcel Dekker.

Stevens, R. B. (ed.) (1974). *Mycology Guidebook*. Seattle & London: University of Washington Press.

Straus, J. (1985). *Industrial Property Protection of Biotechnological Inventions: Analysis of Certain Basic Issues*. Document BIG/281. Geneva: World Intellectual Property Organization.

Tommerup, I. C. & Kidby, D. K. (1979). Preservation of spores of vesicular-arbuscular endophytes by L-drying. *Applied and Environmental Microbiology* **37**, 831–5.

Tuite, J. (1968). Liquid nitrogen storage of fungi sealed in a polyester film. *Mycologia* **60**, 591–4.

Turner, W. B. (1971). *Fungal Metabolites*. London & New York: Academic Press.

Turner, W. B. & Aldridge, D. C. (1983). *Fungal Metabolites* II. London: Academic Press.

Voss, E. G. (ed.) (1983). *International Code of Botanical Nomenclature adopted by the Thirteenth International Botanical Congress, Sydney, August 1981*. Regnum Vegetabile 111. Utrecht & Antwerp: Bohn, Scheltema & Holkema.

Wellman, A. M. & Walden, D. B. (1964). Qualitative and quantitative estimates of viability for some fungi after periods of storage in liquid nitrogen. *Canadian Journal of Microbiology* **10**, 585–93.

WIPO (1980). *Records of the Budapest Diplomatic Conference for the Conclusion of a Treaty on the International Recognition of the Deposit of Microorganisms for the Purposes of Patent Procedure*. WIPO Publication no. 332 (E), p. 119. Geneva: World Intellectual Property Organization.

WIPO (1986). *Report adopted by the 2nd session of the Paris Union Committee of Experts on Biotechnological Inventions and Industrial Property*. WIPO Document BioT/CE/II/3. Geneva: World Intellectual Property Organization.

Wu, L.-C. (1987). Strategies for conservation of genetic resources. In *Cultivating Edible Fungi*, ed. P. J. Wuest, D. J. Royse & R. B. Beelman, pp. 183–211. Amsterdam: Elsevier.

INDEX

201

WITHDRAWN